Yorkshire Chemicals Limited

LEEDS

TECHNICAL INFORMATION BUREAU

Accession No.:   10,838

Date: 14.11.78

Location: Mr. Beanlands

ADDED VALUE
the key to prosperity

ADDED VALUE

the key to prosperity

by E. G. WOOD BCom,MIMC,MBIM

*Director of the Centre for
Innovation and Productivity,
Sheffield City Polytechnic*

BUSINESS BOOKS
COMMUNICA - EUROPA

First published 1978

ISBN 0 220 66349 1

Printed by The Anchor Press Ltd., Tiptree, Essex, and
bound by Mansell (Bookbinders) Ltd., Witham, Essex, for the
publishers, Business Books Limited, 24 Highbury Crescent, London N5

Contents

DEDICATION

To all the writers and speakers who influenced my thinking on the subject of added value. Their efforts have simplified the task described by Dr Samuel Johnson:

'A man will turn over half a library to make one book.'

Preface

About this book

The need for a book on the added value concept arose from my experience of using added value over a period of 20 years, first as a management consultant in industry and commerce, then in research work and also in conducting seminars. Although there have been many articles in newspapers and journals about added value, few books have been devoted to the subject. This book is an attempt to outline in a coherent theme the fundamental concept of added value and its various uses.

Added value is not new. It has been around for nearly 200 years. Tenche Cox, a US Treasury official, devised the idea. Economists have been using the concept for over 100 years in national accounting. Yet it is only in the last 50 years that other people have started to pay attention to it.

First came the managers, using added value as a basis for bonus schemes. But added value has much wider connotations than pay. It is especially useful for measuring company performance, for monitoring manpower productivity and for communications. Over 20 years ago, Peter Drucker, in *The Practice of Management,* wrote about productivity and the concept of what he called 'contributed value'. But still today far too few managers fully understand added value and use it in their business.

Whether they like it or not, managers need to learn more about added value. Why? Because the social climate has changed.

Employees are no longer hired hands. They are being taught to think for themselves. They want to know more about what goes on. But most of them do not understand the nature of business and industry. They need explanations. The added value concept provides a basis for outlining what business is about. The added value statement is a far better way of describing performance than the profit and loss account.

More recently, the accountants have taken up the theme. One of the recommendations of the Accounting Standards Committee is that companies should publish statements of added value. As yet, accountancy textbooks do not mention added value. This is not surprising when accountants themselves have been brought up to account for the profit and the capital yet not for the wealth created. But already, added value has crept into the examination questions of the accountancy bodies. Soon, every accountant will need to know how to calculate added value and how to present information for employees as well as shareholders.

Governments too have woken up to the importance of added value. Cabinet members have started to talk about it. Civil servants also are using the concept more widely. And trade unions are teaching their members what added value means so that they can ask the right questions in negotiations with management.

Now added value has become fashionable, there is a danger that it may be seen as a cure for all ills. But added value is no panacea. It has its limitations. It does not replace the profit motive as a spur for improvement. But it does overcome the emotional connotations of the profit concept. Return on capital ratios are useful for investors, but added value ratios are important for both employees and investors. Profit ratios can vary widely with accounting practices, but added value figures are less readily distorted. For measuring efficiency in the use of resources, the added value concept has advantages over other techniques. It is less distorted by inflation. Most important, it emphasises the fundamental connection between capital investment, manpower productivity and wages.

Added value can be used to measure business performance and manpower productivity. It provides a basis for wage and salary policies and for corporate bonus plans. It is relevant in determining marketing strategies by highlighting the products and markets offering the best added value ratios. It can assist in capital investment appraisal by comparing the added value available from different investments. It provides a new set of business ratios relevant to all parties, employees, investors and customers. Above all, it can facilitate the understanding of accounting information for employees and other users. Thus it can play a part in plans for employee participation.

Added value is the key to prosperity. It measures the creation of wealth and determines the standard of living. It is fundamental to our future welfare. A better understanding of its importance at all levels would help to improve the health and strength of the economy.

One final question. Should it be called added value or value added? It matters little. This book adopts the term 'added value' mainly to avoid confusion with value added tax (VAT). The latter is really a misnomer. It is not a tax on value added but a sales tax. It happens to be collected in stages as the value is added. But the tax is borne by the final customer who adds no value.

Outline of the book

The book has nine chapters, each dealing with a main aspect of added value.

The first chapter, entitled 'Added Value and Its Uses', outlines the fundamental concept of added value. It discusses definitions of added value and describes what happens to added value. The chapter concludes with a list of the main uses of added value.

Chapter 2, 'Added Value and National Accounting', shows how added value is used as the basis of calculating national income and expenditure. It also outlines the *Census of Production* and shows some of the uses of the data therein. In particular the relationship between manpower productivity, expressed as added value per head, and capital investment is discussed.

The third chapter, 'Added Value and Business Performance', starts with a critique of profit as a measure of performance. It goes on to outline the merits of added value for measuring performance. It concludes with a case history, based on the ICI published accounts, comparing the added value and profit methods of assessing performance.

Chapter 4, 'Added Value and Communications', asks and attempts to answer the question: 'What is business about?'. It discusses the wider purposes of industry and commerce as a means of creating wealth for the benefit of customers, employees and the providers of capital. The chapter concludes with some comments on the need for better understanding of the wealth creation process.

The fifth chapter, 'Added Value and Accounting Information', starts by outlining the shortcomings of the conventional profit and loss account. It goes on to discuss the advantages of the value added statement and its use in communications. Finally it outlines various methods of presenting accounting information in added value terms to facilitate understanding.

Chapter 6, 'Added Value and Wages and Salaries', reveals the link between added value and pay. It shows the stability of the ratios in various countries and industries. It discusses the use of added value for pay policy at national level and at company level.

The seventh chapter, 'Added Value and Bonus Schemes', outlines various types of incentive schemes then deals with the principle of added value based bonus schemes. It outlines the factors involved in the design of such schemes. It concludes with some practical points about installing added value based bonus schemes.

Chapter 8, 'Added Value and Business Policy', starts with the relevance of added value to marketing strategy. It goes on to discuss product analysis and pricing policy. The next section deals with capital investment policy. The chapter concludes with a discussion of added value for business ratios.

The last chapter, 'Added Value and Wealth Creation', stresses the need to generate more added value. It discusses the relevance of price increases, innovation and cost reduction. Finally it emphasises the importance of productivity, the creation of more added value from existing resources.

There is also a Bibliography which lists the few books and many articles on the subject.

Acknowledgements

My thanks must go to various people who have written on added value or with whom I have discussed the subject. I am grateful to Douglas Bentley, John Burns, Bernard Cox, John Dickinson, Ron Gilchrist, Brian Jevons, Frank Jones, James Lee, Rene Lee, Sol Margolis, Grange Moore, Dudley Newton, Roy Pickering, Russell Smallwood, Geoff Smith, Alan Thompson, John Wellens and Duncan Wood and others to numerous to mention. The book is the better for their contributions. But the responsibility for any shortcomings is entirely mine. The task of writing and rewriting required the patience and understanding of my wife and family. Last but not least, my secretary, Sandra Hibbard, deciphered my scribble and corrected various typescripts with accuracy and efficiency.

ADDED VALUE and its uses

1

1.1 What is added value?

Added value is a form of wealth. But not all forms of wealth are added value. Added value is not the kind of wealth that occurs as natural resources. Things like land, minerals, metals, coal, oil, timber, water and similar sorts of wealth are provided by nature. Added value is the kind of wealth generated by the efforts and ingenuity of mankind. The word wealth is commonly used to describe opulence, riches or large possessions. Such wealth may have come partly from natural resources and partly from the wealth created by people, i.e. added value. Much of the added value created is consumed soon after it is generated. But some of it is accumulated in the form of buildings and capital equipment.

At the primitive level, a man goes into a forest and cuts down a tree. He converts it into a house, furniture and other articles for his own use. In doing so, he 'adds value' to the raw material provided by nature. In our complex industrial society, a manufacturing business buys raw materials, components, fuel and various services. It converts these into products which can be sold for more than the cost of the raw materials and other purchases. In doing so, the business 'adds value' to the materials by the processes of production. Similarly, a farmer generates wealth by growing crops and breeding animals, then selling them for more than the cost of the seeds, fertiliser, foodstuffs and other materials used.

A retail business buys goods, stores them, displays them and sells them. Thus it 'adds value' to the goods by providing a service for which customers are willing to pay. Similarly, a restaurant 'adds value' to the raw food by cooking it and serving it to customers who pay more than if they had bought the food and cooked it themselves. A transport business 'adds value' by moving goods from one place to another. It provides a service for which customers are willing to pay. By changing the form or location of products, these businesses 'add value' to the materials.

Added value may be generated even when little or no material is involved. A hairdresser 'adds value' by offering a service for which customers pay. The costs of the materials are small. An entertainer provides a service. The customer pays for the 'added value' of the enjoyable experience. A doctor also provides services with little material cost. People pay directly or indirectly for the 'added value' of good health.

The common factor in most of these examples is that the customer pays for the product or service. The gap between what the customer pays and what the manufacturer or supplier has to pay for raw materials, and other bought-in items, is the added value generated. In the case of the primitive man making his own furniture from 'free' raw materials, there is no monetary transaction. But the lack of a monetary measure does not mean that no wealth has been created. Similarly, the modern do-it-yourself addict may 'add value' to his house by enhancing its function or aesthetic appeal. In practice it is easier to measure the added value of a manufacturing business than that of a handyman householder. One of the problems of the modern economy is that it measures only the activities for which money is exchanged. The housewife and the housekeeper do similar work but the services of the unpaid housewife are not included in the national income.

It should be noted that the measure of added value is not the effort that has gone into the activity. Added value is determined by the satisfaction of the customer, not by the work of the producer. Manufacturing quill pens might need considerable effort on the part of the workers and managers. But if nobody wants to buy quill pens, no added value has been generated. When a business or industry has to be subsidised to keep it going, the cost of the subsidy, along with the cost of materials and purchased services, must be deducted from the sales revenue to calculate the added value.

What about the non-marketed activities in a modern society? Does a hospital or a school generate wealth? When a doctor enhances the health of a patient, both the individual and society benefit from the service. When a teacher inculcates knowledge and develops skills in the pupil, both the individual and society benefit from the service.

In this sense, medical and educational services can be said to generate wealth. Of course, it is not wealth in the sense of a manufactured product or a material possession. Nevertheless, these services help to raise the standard of living. But when these services are financed by taxation, it is not easy to measure the 'added value'. If there is no 'profit' or 'loss' it is difficult to know whether the added value created is greater or less than the sum total of the costs involved. It is easier to discuss added value in terms of the more straightforward examples of industry and commerce.

Added value is simply a measure of the wealth created by a business or an industry. It differs fundamentally from sales revenue because it excludes the wealth created by the suppliers of the businesses. Thus, added value is a measure of the net output rather than the gross output of a factory, company, industry, or even a country. Added value overcomes the problem of double counting when the outputs of several businesses are added together. Making a garment starts with the farmer who grows the wool or cotton. The raw material passes through the hands of the spinner, the weaver, the dyer and printer, the clothing manufacturer, the retailer and all the intermediate transport operators. If the sales turnovers of all these businesses are added together, the total includes much double counting. If the added value figures for all the businesses are added together, there is little or no double counting. The sum total of their added value figures represents the wealth created by the collective efforts of the businesses.

However, there is no universally accepted definition of added value. There is general agreement about the concept of added value, i.e. it measures the wealth created by a business or industry. But there are some differences of opinion about the precise definition. Indeed, there are good arguments for having slight differences in definition for different purposes. Fortunately, the differences in practice are usually small in magnitude. Nevertheless, it is important to understand the basic principles. How, therefore, should added value be calculated?

1.2 How to calculate added value

At first sight, added value can be measured very easily in terms of the difference between sales and purchases. The gap between what a business charges for its products and what it pays for the raw materials and other purchases represents the value added to the materials by the processes of production. The principle is illustrated in Figure 1.1. In practice it can be more complex.

First, there is the question of stock changes. In any one year, or

Figure 1.1 Calculating added value - stage 1

	£
Sales, less discounts	960,000
Purchases of materials and services	520,000
Added value (approximately)	440,000

month, a business may buy more raw materials, etc., than it actually uses. So the stock of raw materials, etc., will be greater at the end of the period than at the beginning. Conversely, if the purchases are less than the amount consumed, the stock will fall. If stock changes were ignored, then two periods with identical figures for sales turnover could show different added value figures for no other reason than the change in stocks. So it is sensible to adjust for stock changes, in much the same way that adjustments are made in financial accounting in order to calculate a profit figure. The calculation should also take account of writing off any stock of obsolete material.

Figure 1.2 Calculating added value - stage 2

	£	£
Sales, less discounts		960,000
Purchases of materials and services		520,000
Opening stock of raw materials, etc.	80,000	
Closing stock of raw materials, etc.	100,000	
Increase in stock	20,000	
Purchases, less increase in stock		500,000
Added value (after partial stock adjustment)		460,000

The method of adjusting for changes in stocks of raw materials, etc., is shown in Figure 1.2. But in the same way that the stocks of raw material, etc., may fluctuate, so the stocks of work-in-progress and finished goods may change. Thus, the sales figure may represent more than, or less than, the value of the manufactured output. In financial accounting, adjustment is made for such stock changes when calculating the profit. But should such adjustments be made when calculating added value? In theory, it can be argued that no value is added until the goods are sold. In practice, it is reasonable to assume that added value is generated when the goods are manufactured. It is true that until the stocks of finished goods are actually sold, there is no certainty that they will fetch a price higher than the cost of the materials. But in a sound, healthy business, most of the finished stock will eventually be sold at the normal price. In financial accounting, finished goods are usually valued at less than the selling price but at more than the cost of raw materials, etc. In calculating added value, the figures derived for financial accounting can be used for the sake of simplicity and comparability. After all, it is only the change in stock that matters, not the level of total stock. Of course, the opening and closing stocks should be valued on a similar basis. The method of adjusting for changes in stocks of finished goods and work-in-progress is shown in Figure 1.3, which also incorporates the changes in stocks of raw materials, etc., shown previously in Figure 1.2.

It should be noted that the increase in stock of finished goods is added to the sales figure whereas the increase in stock of raw material is deducted from the purchases. It is all too easy for the uninitiated to add the increase (or subtract a decrease) in both cases. But commonsense indicates that if the closing stock of raw material is higher than the opening stock, the amount of materials actually used

Figure 1.3 Calculating added value - stage 3

	£	£
Sales, less discounts		960,000
Opening stock of finished goods and work-in-progress	80,000	
Closing stock of finished goods and work-in-progress	120,000	
Increase in stock	40,000	
Gross output = Sales, plus increase in stock		1,000,000
Purchase of materials and services		520,000
Opening stock of raw materials, etc.	80,000	
Closing stock of raw materials, etc.	100,000	
Increase in stock	20,000	
Purchases, less increase in stock		500,000
Added value (after full stock adjustment)		500,000

must be less than the amount purchased. In calculating added value, it is the usage of materials that matters, not the purchases. Conversely, if the closing stock of finished goods is higher than the opening stock, it is commonsense to add the increase to the sales figure to give the gross output of the business.

An easier way of arriving at the added value figure is to calculate the stock change in one step rather than two. In the example shown in Figure 1.3, the total opening stock is £80,000 of raw material, etc., plus £80,000 of finished goods and work-in-progress, making £160,000 all told. The total closing stock is £100,000 of raw materials, etc., plus £120,000 of finished goods and work-in-progress, making £220,000 all told. So the increase in total stock is £60,000. This figure can be added to sales to give £1,020,000. When the purchases of £520,000 are deducted, it gives the same added value figure of £500,000.

Alternatively, the increase in stock of £60,000 can be deducted from the purchases of £520,000 to give £460,000. When this figure

Figure 1.4   Ratio of added value to gross output

	Year 1 £000	Year 2 £000	Year 3 £000
Sales	960	980	1040
Finished goods:			
Opening stock	80	120	140
Closing stock	120	140	100
Stock change	+40	+20	−40
Gross output	1000	1000	1000
Purchases	520	480	520
Raw materials:			
Opening stock	80	100	80
Closing stock	100	80	100
Stock change	+20	−20	+20
Purchases, adjusted for stock change	500	500	500
Added value	500	500	500
Ratios:	%	%	%
Purchases/sales	54	49	50
Added value/sales	52	51	48
Added value/gross output	50	50	50

is deducted from the £960,000 of sales it gives the same added value figure of £500,000. The answer is the same by all three methods. So which method is best?

For speed of calculation, the one-step methods are easier. But the two-stage method is worth the little extra trouble. Why? Because it provides a figure of gross output (sales, adjusted for stock changes). The ratio of added value to gross output should not be significantly affected by changes in stock levels. By contrast, the ratio of purchases to sales can vary simply *because* stock levels change. The example shown in Figure 1.4 illustrates the point. In each of the years the ratio of purchases to sales varies significantly. Yet the ratio of added value to gross output is constant. The two-stage method facilitates the calculation of this useful ratio. However, if this ratio is not required, both of the one-step methods of calculating added value will give a correct figure. Of course, if the stock figures do not distinguish between raw materials and work-in-progress and finished goods, the one-step method must be used.

The examples quoted above assume that the increases in the values of stocks are associated with increases in the physical quantity of stocks. However, increases in the value of stocks can also be caused by inflation. A given quantity of goods at the end of the year may have a higher monetary value than the same quantity at the beginning of the year. Financial accounting based on historical costs shows the increase in stock value as profit. Similarly, a calculation of added value based on historical costs would show an inflationary increase in stock value as part of the added value. Clearly, in calculating both profit and added value we ought to use some form of inflation accounting to adjust for stock appreciation. This problem is too complex to discuss here. Suffice it to say that, using historical costs, added value figures are distorted far less than profit figures.

The effect of inflation on profit and added value is illustrated in Figure 1.5. The first column shows the same data as Figure 1.3, with the addition of employment costs, depreciation and profit. The second column shows the effect of inflation, at the rate of 1 per cent per month, on sales, purchases and employment costs. In practice, inflation affects these three variables at different rates per month. But that does not invalidate the logic of the argument.

The effect of the inflation in stock values is to increase the added value by 10.6 per cent. But the profit before interest and depreciation is increased by 18 per cent. And the profit after depreciation could be up by 36 per cent. Of course different results would emerge with different assumptions about the rate of inflation, the levels of stocks in relation to sales and purchases, the ratio of added value to gross output, the ratio of employment costs to added value and the

Figure 1.5 Effect of inflation

	Inflation rate		Percentage increase
	Nil	1% per month	
	£	£	
Sales, less discounts	960	1015	5.7
Opening stock	160	160	–
Closing stock	220	248	12.7
Increase in stock	60	88	46.7
Sales plus stock increase	1020	1103	8.1
Purchases	520	550	5.7
Added value	500	553	10.6
Employment costs	300	317	5.7
Profit before interest and depreciation	200	236	18.0
Depreciation	100	100	–
Profit before interest	100	136	36.0

figure for depreciation. But whatever the assumptions, *inflation distorts profit far more than it distorts added value.*

The adjustment for stock changes is not the only problem in calculating added value. It is also important to define what is meant by sales turnover. In many firms there is no difficulty. But some companies receive royalties on patents, dividends from associated companies, income from investments, rent from lease of buildings and other miscellaneous items. In practice, these items are usually small in relation to sales turnover from manufacturing and trading activities. So it makes little difference whether they are included or excluded. But in using added value to compare one year with another or one company with another, care should be taken if these items are significant.

Another problem in calculating added value is that of deciding exactly which purchases should be subtracted from sales. There is no doubt that the cost of raw materials, bought-in components, packaging materials and any items which form part of the finished product must be deducted, after stock change adjustment. Similarly, fuel costs should be deducted irrespective of whether the fuel is used mainly for processing, as in steel making or textile finishing, or purely for space heating and lighting. Other items like the consumable stores and loose tools used in an engineering factory should also be deducted. The cost of materials used for repairing and maintaining the plant, equipment and buildings should be subtracted, but not the cost of materials used to build *new* plant and equipment or new buildings. The latter items would form part of the capital expenditure which could be funded either out of retained profits (part of the added value) or by money borrowed from a bank or other sources. Finally, the costs of stationery and office materials should also be deducted. In practice, some of these items are very small and it makes very little difference whether they are included or excluded.

The tricky items are not so much the purchases of materials but the purchases of services. The costs of transport, whether for raw materials in, or for finished goods out, should be deducted. In some cases, transport charges are not identified separately, e.g. the price paid for raw materials may include the transport costs. But if the business operates its own transport fleet, whether for materials inwards or finished goods outwards, then the costs of fuel and spare parts should be deducted. The other costs of transport, namely ways and depreciation, come out of the added value of the whole enterprise. Care must then be taken in comparing two similar businesses, one with its own transport fleet, the other using external transport services. Fortunately, in most manufacturing industries, the costs of transport are small in relation to the added value, usually less than 1 per cent.

In calculating the added value the costs of advertising should be deducted. These may include fees paid to advertising agencies, newspapers, television, etc. However, if the enterprise has its own advertising department, then the costs of materials used, e.g. for display stands, should be deducted. But the wages, salaries and other employment costs of the advertising department are part of the added value. Any payments made for legal services, the audit fee and other professional fees should be deducted. But the salaries of the company's own lawyers, internal auditors or other experts, are part of added value. Commission paid to agents should be deducted. But commission paid to the company's own sales representatives is part of the added value. Discounts offered to, and taken by, customers should be deducted. But discounts received from suppliers are part of the added value.

The general rule is to deduct those costs that represent the output of suppliers of goods and services. Thus the costs of postage and telephones should be deducted. These are services purchased by the business. Other service costs which would normally be deducted are bank charges (but not the interest charges on borrowed money), commercial insurance premiums and subscriptions to trade associations.

9

One of the debatable items is travelling expenses. Where these are simply reimbursement of actual costs of rail fares, air fares, hotels, etc., then all such expenses should be deducted. But if some of the expenses represent a benefit to the employees, they are part of the added value. For example, some employees may benefit from company cars used partly for private purposes. In practice, it may not be worthwhile trying to sort out the difference between expenses paid out for services and expenses paid as benefits to employees.

Another debatable item is rates. Some people argue that rates are a payment for services rendered by local authorities. If so, rates should be deducted when calculating added value. Other people argue that rates are a form of tax which should be included in added value. However, there is little doubt that the charge for water rates is a purchased service. Again, in practice, the difference between including or excluding rates from added value may be very small. Nevertheless, comparisons should be done on a consistent basis.

Then there is the question of rent and hire charges. One school of thought treats these as service costs to be deducted. Thus they are part of the added value of the owners of the buildings or plant. However, this procedure makes for difficulties in comparing, say, one company that owns all its buildings and plant with another company that rents its buildings and hires all or part of its plant. The company that hires all or part of its plant will show no depreciation or, at least, a smaller amount than if it bought all its plant.

Depreciation is normally part of added value. However, some authorities like to distinguish between *gross* added value and *net* added value. The former includes depreciation and possibly hire charges and rent. The latter excludes depreciation (based on historical costs) and possibly stock appreciation. The concept of net added value has certain uses. But for most practical purposes, added value should include depreciation. In calculating gross added value it matters not whether the depreciation is calculated on historical costs or replacement costs. To the extent that depreciation is increased, profit is reduced. The added value remains the same, irrespective of changes in depreciation policy.

Finally, there is the problem of taxes. Does added value include value added tax and other indirect taxes paid by consumers? Most manufacturing companies pay VAT to their suppliers. But the VAT they pay does not form part of their costs. They can deduct it from the VAT which they collect from their customers. Similarly, many of their customers are registered for VAT and can pass on the tax. Only the ultimate consumer, who cannot pass on the VAT, bears the tax. Economists distinguish between 'added value at market price' and 'added value at factor cost'. The former includes VAT and other taxes. The latter excludes taxes. It represents only the cost of the factors of production, including the profit element. For many purposes in using the added value concept, VAT can be ignored, especially if the rate is the same for all goods and services. But it can be argued that the government is syphoning off a percentage of the added value out of the price that customers are prepared to pay.

Similarly, customs and excise duty on petrol and oil, tobacco, beer, wines and spirits, are part of the 'added value at market price' but not part of 'added value at factor cost'. These special taxes, like VAT, skim off part of the price that customers pay. But if taxes were reduced, prices might fall, perhaps in line with the reduction in taxes. Certainly, when indirect taxes are raised, most of the increase is borne by the ultimate consumer and not by the supplier. Such taxes are a special form of cost that should be deducted when calculating the added value for practical purposes.

In principle, there are various methods of calculating added value. In practice, the differences are fairly small. There is no argument that added value must include all wages, salaries and other employment costs. It must also include depreciation, interest charges and profits before corporation tax. Most of the debatable points have only a minor effect on the added value figure. There may be a case for having slightly different definitions depending on the purpose for which the added value figures are needed. The vital point is that, in any comparisons between organisations, or any comparisons within one organisation over a period of time, the same definition should be used throughout.

To illustrate the principles and practice of calculating added value, Figure 1.6 shows data based on a real company. The actual figures have been rounded off to simplify the arithmetic. The added value

Figure 1.6 Calculating added value in practice

	£	£
Sales turnover, after discounts	2,480,000	
Agents' commission	20,000	
Sales less commission	2,460,000	
Opening stock, finished goods and work-in-progress	200,000	
Closing stock, finished goods and work-in-progress	240,000	
Change in stock	+40,000	
Gross output		2,500,000

(Continued overleaf)

Purchases of raw materials	1,000,000	
Opening stock, raw materials	85,000	
Closing stock, raw materials	108,000	
Change in stock	+23,000	
Raw materials used		977,000
Other materials:		
Consumable stores	15,000	
Electricity	5,000	
Fuel oil	9,000	
Packing material	20,000	
Repairs to buildings	10,000	
Repairs to plant	24,000	
Stationery	8,000	
Water	2,000	
Total other materials		93,000
Total all materials		1,070,000
Purchased services:		
Accountancy fees	4,000	
Advertising	28,000	
Bank charges, excluding interest	2,000	
Carriage	22,000	
Insurance	11,000	
Legal fees	1,000	
Miscellaneous expenses	7,000	
Motor vehicle expenses	8,000	
Plant hire	1,000	
Postage	2,000	
Rates	12,000	
Rent	3,000	
Subscriptions	1,000	
Telephone	12,000	
Travelling expenses	16,000	
Total purchased services		130,000
Total materials plus services		1,200,000
Added value		1,300,000

of £1,300,000 represents 52 per cent of the gross output. If all the debatable items like plant hire, rent and rates were to be included, the added value would then be £1,316,000. The ratio of added value to gross output would then be 52.6 per cent. So, in practice, the difference is very small.

Figure 1.7 shows the disposal of the added value classified into four groups. These are employment costs, financial charges, depreciation and net profit before tax. The depreciation could be deducted to give a *net* added value figure of £1,233,000. The ratio of net added value to gross output would be 49 per cent. The change is significant but not great. However, in a more capital-intensive business the difference could be much greater.

Figure 1.7 Disposal of added value

	£	£
Employment costs:		
Direct wages	350,000	
Indirect wages	180,000	
Staff salaries	210,000	
Directors' remuneration	45,000	
Holiday pay	55,000	
National insurance	75,000	
Pension contributions	50,000	
Canteen subsidy	5,000	
Other employee benefits	2,000	
Total employment costs		972,000
Financial charges:		
Bank interest	18,000	
Hire purchase interest	6,000	
Mortgage interest	2,000	
Total financial charges		26,000
Depreciation charges:		
Depreciation, plant	62,000	
Depreciation, vehicles	15,000	
Total depreciation		77,000
Net profit before tax		225,000
Added value		1,300,000

1.3   What happens to added value?

Added value is a measure of the wealth generated by the collective effort of those who work in a business or industry — the employees and owner-managers — and those who provide the capital — the investors. The added value is used to pay those who contribute to its creation. In most companies, the biggest slice of the added value goes to the employees in the form of wages, salaries, bonuses, holiday pay, the employer's contribution to pensions and national insurance, and all other forms of remuneration or employment cost. Another slice goes to those who provide the capital. This is paid either in the form of interest or dividends. However, before the shareholder can receive a dividend, the government demands a slice of the added value in the form of tax on profits, i.e. corporation tax. It could be said that the company is thus paying for services provided by the government. But this argument is not really acceptable. Companies that incur losses do not pay corporation tax. But they still enjoy or suffer the government services. The final slice of added value is retained in the business in the form of depreciation and retained profit. Strictly speaking, the depreciation would be deducted in order to calculate the net profit which includes dividends and corporation tax. In effect depreciation can be regarded as a charge for the ownership and use of the assets. But unlike a charge for leasing or hire of assets, depreciation is not a flow of cash out of the business.

Thus, in sharing out the added value, there are four major slices:
1   Wages, salaries and other employment costs — for employees.
2   Dividends and interest on loans — for investors.
3   Taxes on profit — for government.
4   Depreciation and retained profit — to create more wealth.

Many British companies are now publishing an added value statement in their annual report and accounts. The form of statement varies slightly from one company to another. Most of them follow the format recommended by the Accounting Standards Committee in their discussion paper, *The Corporate Report*. An illustration of an added value statement is given in Figure 1.8.

The added value statement shows how the wealth created by the enterprise has been used to pay those contributing to its creation. It puts profit into perspective as only a relatively small part of the wealth generated by the business. Out of the added value comes wages, taxes, interest and dividends, and funds for further investment. The interdependence of manpower and capital is made more apparent by the added value statement.

The monetary values in Figure 1.8 represent a fairly small manufacturing business. But even if the business were 10 or even 100 times the size, the relationship between total wages, taxes, interest

Figure 1.8 Added value statement

XYZ Manufacturing Company - ADDED VALUE STATEMENT

		£
	Sales (adjusted for stock change)	1,000,000
Less:	Materials and services used	500,000
	ADDED VALUE	500,000
Disposal of added value:		
	Employees:	
	Wages, pension and national insurance contributions, etc.	350,000
	Government:	
	Corporation Tax payable	30,000
	Providers of capital:	
	Interest on loans	15,000
	Dividends to shareholders	30,000
	Re-invested in the business:	
	Depreciation	35,000
	Retained profits	40,000
	ADDED VALUE	500,000

and dividends, depreciation and retained profits would be similar. The lion's share of the added value goes to the employees. The providers of capital receive a much smaller share, often little more or even less than the government takes in taxes. In theory, the retained profits belong to the shareholders. In practice, they are re-invested in the business to generate more added value which in turn benefits employees and the government more than shareholders.

There are other ways of looking at the distribution of the added value. Within the employment costs is the income tax paid by employees. Thus it can be argued that the government takes an even bigger slice of the added value than is represented in Figure 1.8. Similarly, the people receiving the interest charges and dividends might be liable to pay tax on such income. Also, if VAT and customs and excise duties were included in the sales and in the added value, the proportion going to the government would be even higher. Finally, the national insurance contributions paid by employees and employers goes to the government but there is no precise link between the contributions and the benefits. Part of the contributions can be regarded as another form of tax.

Conversely, there may be some payments from the government to a company, e.g. grants for investment, subsidies for employment, etc. Some companies have adopted the practice of showing corporation tax, less grants, as a net figure in the added value statement. It would be better if the grants were shown separately. Similarly, employment costs have been shown net after any subsidies. It would be better if the full amount of the subsidies were disclosed. The grants and allowances should not be regarded as wealth generated by the company but rather as wealth generated elsewhere and transferred to the company.

A distinction must be drawn between grants for investment and the capital allowances offering relief of corporation tax against the cost of capital expenditure. Capital allowances simply reduce the tax that would have been paid out of the added value. But investment grants are direct subsidies irrespective of any tax liability. In theory, the distinction is fundamental. In practice the effects may be very similar in a profitable company. What matters is whether an organisation is generating enough added value not only to pay the employees and any taxes but also to finance its own investment programme. Subsidies may be hiding excessive wages and over-manning. Grants may be distorting the investment strategy.

In simple terms, added value covers wages, salaries and other employment costs, depreciation, interest charges and profit before tax. It is sometimes easier to calculate the added value of a company by adding these components together than by the method of subtracting the purchases from the sales and making all the various adjustments. Indeed, the additive method can be used to verify the arithmetic of the subtractive method, or vice versa. However, there is then the danger of assuming that added value arises as a consequence of paying out wages and interest charges, setting aside depreciation and generating profit. Nothing could be further from the truth. It is the creation of added value that permits the payment of wages and interest charges and that provides a fund for depreciation and profit. Increasing the wage bill does not increase the added value. It simply reduces the profit. Conversely, reducing the wage bill does not reduce the added value. It simply increases the profit.

In theory, only a monopoly organisation can pass on wage rises in the form of higher prices. In doing so, it appears to increase its added value. But its customers will find that their added value will fall, unless they too can pass on the increase in costs. If many organisations pass on the increases, the result is inflation; higher prices for the same quantity of goods. In a competitive market, added value is determined by the customer, not by the supplier. But in practice, few markets are truly competitive. Modern business tries to create monopolies or at least conditions in which wage increases can be passed on as price increases.

This simple truth is the basis of the wage-price spiral. If a company is faced with an increase in its total wage and salary bill, with no corresponding increase in real output, it will try to put up its prices to maintain its profit. So price rises follow wage increases. Conversely, if wages and salaries are stable and companies try to increase their profits by raising prices, employees will seek wage increases to compensate for higher prices. Thus wage increases follow price rises and price rises follow wage increases. Added value is the link between wages and prices. The three factors are inextricably interrelated.

1.4   The main uses of added value

The concept of added value has many uses. It provides a method of measuring the net output of a business, industry, or country. It is the basis of national accounting. As a measure of output, added value is more useful than other measures such as physical quantities, sales turnover, or profit. But it does not replace them.

Because it provides a measure of output, added value can be used to measure manpower productivity in terms of added value per employee. Within one company, comparisons of added value per employee can be made over successive months or years. If necessary, adjustment can be made to compensate for inflation to see whether increases in added value per employee represent real growth or simply growth in monetary terms. Comparisons of these ratios between companies or industries must be treated with more caution. The differences may be associated with differences in the level of capital investment per employee or with the effects of monopoly profits.

Added value can also be related to the amount of capital employed thus giving a measure of capital productivity. In practice, there are problems of measuring the amount of capital employed. Balance sheet figures may vary with accounting procedures. But the principle of relating output to capital employed is still valid. Again, comparisons within one organisation over periods of time are easier than comparisons between different organisations and industries. Despite these problems, added value is still a more reliable measure for productivity comparison than either sales or profit.

Because added value represents the wealth created by an enterprise it provides a sound basis for explaining business activities in simple terms. It can facilitate the provision of accounting information for employees and others. The added value statement is far easier to understand than the conventional profit and loss account. Explaining a business in terms of added value overcomes the emotional connotations of the word profit. Instead of trying to justify the need for profit as a reward for capitalists, the added value concept justifies the need to generate wealth which can be shared out between employees and investors (and the government, in taxes). Nowadays many employees are also investors indirectly through pension funds, insurance companies and taxation.

Once the concept of added value is understood by employees at all levels, the traditional conflict between labour and capital can be reduced, if not eliminated. In turn this can lead to genuine participation in decision-making for the benefit of all. There may still be room for argument about how the added value is to be shared out. There should be no doubt about the need to cooperate in order to create more added value.

Added value can be used as a basis for wage and salary policies. The index of added value per employee is the vital figure. Why? Because it sets a limit to the average wage per employee. No company can pay out more in wages per employee than it is generating in added value per employee (unless it is subsidised). The higher the added value per employee, the higher can be the average wage per employee. Of course, the cost of capital, including depreciation, interest charges and dividends, must also be covered by the added value. So a high ratio of added value per head may indicate high investment per head. But high investment per head does not guarantee high added value per head. The creation of added value depends not on the level of capital expenditure but on good marketing strategy, sound investment policy, effective management and employee cooperation to maximise the added value per employee. The only way to raise real wages is to generate more added value per person.

Thus added value can provide a basis for an annual negotiation of wage and salary levels. Some companies have used added value as the basis for group bonus schemes. The original added value based bonus scheme, the Rucker plan, was introduced in the USA in the 1930's. There are now many variations on the basic theme. Such schemes can help to foster cooperation to increase output and to reduce material costs. The link between output and wages becomes patently obvious in an added value based bonus scheme. However, care is required in the design and installation of such schemes. Success depends more on the use of added value to improve communications and cooperation than on any mechanistic link between wages and output.

Another use of added value is for marketing strategy. Most businesses have more than one product. Normally, the ratio of added value to sales revenue varies from one product to another. More important, the ratio of added value to the limiting factors of production, e.g. man hours or machine hours, varies from one product to another. By concentrating on the products with the highest ratios of added value to the limiting factor, the business will generate more added value from the same resources. It will then be able to pay higher wages and better dividends. Sometimes the same product may generate more added value in a different market, e.g. by exporting. A main objective of marketing strategy should be to generate more added value by developing better products and better markets.

Added value can also be used as a basis for investment policy. The conventional practice when making decisions about capital expenditure is to estimate the expected profit from the new equipment and then calculate the return on the investment. But such calculations treat labour as a factor of production, a cost to be minimised. Instead capital expenditure decisions can be based on the added value that should be generated from the investment. The objective is not to maximise profit at the expense of wages but to maximise added value so that both wages and profit can be high.

Finally, added value provides a new set of business ratios relevant to all parties. Whereas the ratio of profit to capital is of interest mainly to the investor, added value ratios are important for both employees and investors. At the same time customers, suppliers and governments want to know about business activities. The narrow concept of profit gives a limited view. The broad concept of added value presents a wider perspective. Added value may not immediately replace the profit motive as a spur to improvement. But it provides a better way of describing performance.

To sum up, the uses of added value can be grouped into four categories as follows:

1   *For measuring output*
    a   Basis of national accounting.
    b   Measuring business performance.
    c   Measuring the productivity of manpower and capital.

2   *For communication*
    a   Explaining what business is about.
    b   Presenting accounting information.
    c   Basis for employee participation.

3   *For rewarding employees*
    a   Basis for wage and salary policy.
    b   Basis for group bonus schemes.

4   *For business policy*
    a   Marketing strategy.
    b   Capital investment policy.
    c   Business ratios.

19

2 ADDED VALUE

and national accounting

2.1 National income and expenditure

The national income is a measure of the monetary value of goods
and services available to the nation from economic activity. It can
be derived in three ways:

1   By adding together all *incomes* from economic activity, i.e.
     incomes from employment and incomes from 'profit' including
     interest and dividends.
2   By adding together all the *expenditures* of the community,
     distinguishing between expenditure on consumption and
     expenditure on investment.
3   By adding together the values of the *net outputs* of all industries
     and commercial activities.

The three methods should produce the same result. The three
measures are calculated each year by the Central Statistical Office
and published in the *'Blue Book'* of *National Income and Expenditure.*
The *national income* includes incomes from abroad as well as incomes
from economic activities in the UK. But it does not include payments
to non-residents. The national income covers only 'factor incomes',
i.e. the income of the factors of production — manpower and capital.
It does not include 'transfer incomes', e.g. pensions, family allowances, grants, etc., which are paid from taxes on factor incomes.

The calculation of *national expenditure* starts with consumers'
expenditure plus the current expenditure by public authorities on
goods and services. To this is added fixed capital formation, i.e.

expenditure on fixed assets such as buildings, equipment and vehicles but not including expenditure on repairs and maintenance. The value of any physical increase in stocks is added next, to give the total domestic expenditure. Then exports and property income from abroad are added. Next, imports and property income paid abroad are deducted. Then taxes on expenditure are deducted. Finally, any subsidies are added to give the Gross National Product or GNP.

The third method of calculation is essentially a sum of the *added value* at each stage of production. This method was devised in 1790 when the first USA census was planned. A treasury official, Cox,

Figure 2.1 Gross domestic product by industry, £million*

Industry or activity	1965	1970	1975
Agriculture, forestry & fishing	1,062	1,344	2,959
Mining & quarrying	714	746	1,684
Manufacturing	10,821	14,833	30,129
Construction	2,164	2,889	6,539
Gas, electricity & water	1,012	1,407	2,995
Transport	1,984	2,708	5,772
Communication	646	1,045	2,809
Distributive trades	3,658	4,680	10,188
Insurance, banking, etc.	2,092	3,380	7,727
Rent	1,395	2,411	5,535
Other services	3,661	5,622	10,454
Public administration & defence	1,805	2,761	7,107
Public health services	613	923	2,816
Local authority education services	817	1,333	4,338
TOTAL	32,444	46,082	101,052
Income from employment	21,292	30,425	68,181
Gross profits and other trading income	11,152	15,657	32,871
Less:			
Stock appreciation	− 318	−1,162	−5,203
Adjustment for financial services	− 905	−1,340	−3,623
Residual error	− 11	− 91	+ 920
GROSS DOMESTIC PRODUCT AT FACTOR COST	31,221	43,489	93,146
GDP AT 1970 PRICES	38,582	43,489	47,453

*Source: *National Income and Expenditure 1965-75, Table 3.1*

realised that if the sales values of all enterprises were totalled, there
would be double counting. He therefore suggested that they should
submit the value of their sales, less the value of everything they
bought. This residual value, or added value, could then be totalled,
company by company, industry by industry, to give the true output.
The added value concept is now used by every major country as a
basis for measuring national output.

Figure 2.1 shows the composition of the Gross Domestic Product
for the UK for 1965, 1970 and 1975. (GDP is the GNP less net
property income from abroad.) The monetary figures have risen
considerably, partly owing to inflation but partly as a result of
genuine growth in the economy. To set the data in perspective,

Figure 2.2 Gross domestic product: percentages by industry

Industry or activity	1965	1970	1975
Agriculture, forestry & fishing	3.4	3.1	3.2
Mining & quarrying	2.3	1.7	1.8
Manufacturing	34.7	34.1	32.4
Construction	6.9	6.7	7.0
Gas, electricity & water	3.2	3.2	3.2
Transport	6.4	6.2	6.2
Communication	2.0	2.4	3.0
Distributive trades	11.7	10.8	11.0
Insurance, banking, etc.	6.7	7.8	8.3
Rent	4.5	5.5	5.9
Other services	11.7	12.9	11.2
Public administration and defence	5.8	6.4	7.6
Public health services	2.0	2.1	3.0
Local authority education services	2.6	3.1	4.7
TOTAL	103.9	106.0	108.5
Income from employment	68.2	70.0	73.2
Gross profits and other trading income	35.7	36.0	35.3
Less:			
Stock appreciation	-1.0	-2.7	-5.6
Adjustment for financial services	-2.9	-3.1	-3.9
Residual error	-	-0.2	+1.0
GROSS DOMESTIC PRODUCT AT FACTOR COST	100.0	100.0	100.0

Figure 2.2 shows the figure for each industry expressed as a percentage of the total GDP. It can be seen that manufacturing industry contributed a smaller proportion of the GDP in 1975 than in 1965. But it still remained the largest single sector.

The figures for agriculture, mining, industry, transport, communications, distribution, banking, etc., are derived by adding the wages and salaries, the employers' contributions, income from self-employment and gross profits including depreciation and appreciation in the value of stocks. Thus they represent the added value generated by each sector, including the appreciation in the value of stocks which was fairly small until the rapid inflation in recent years.

However, the figures for public administration, health and education are derived by adding together wages, salaries and employers' contributions. It is debatable whether the figures can be regarded as the added value generated by these sectors. There is no easy way of telling whether customers would pay these sums in a market economy. The proportion of the total GDP accounted for by these non-commercial activities increased from 10.4 per cent in 1965 to 15.3 per cent in 1975. So it becomes increasingly important to try to obtain a better measure of their contribution.

The *Blue Book* gives more details of the national income and expenditure in terms of personal incomes, company profits and expenditure by public corporations, central government and local authorities. It also gives details of fixed capital formation and stocks of fixed capital. However, most of the figures are highly aggregated, so other government statistics must be used for more detailed investigation. One of the most useful sources of information is the *Census of Production.*

2.2   The Census of Production

The Business Statistics Office collects data from companies in all the manufacturing industries. Summary figures are published in a series of reports. At one time the Census data emerged too late to be of more than academic use. But thanks to computers, the summary now emerges in provisional form within a year. For each of the 150 industries covered by the Census, a mass of information is provided about sales, purchases, numbers of people employed, wages and salaries, capital expenditure, stocks and so on. The 150 industries are grouped together into 20 'Orders' as defined in the *Standard Industrial Classification.*

One advantage of using Census data as a source of information, rather than company accounts, is that the Census gives figures for sales, added value, employment and wages. Another advantage is that, with the growth of large public companies by takeovers and mergers into a diverse range of industries, the annual accounts of public companies give insufficient information to identify what is happening in the various main sectors of industry.

The *Census of Production* collects information at 'establishment' level. An establishment is the smallest unit which can provide the information required for the Census. Usually the principal activities carried on in an establishment fall within a single heading of the *Standard Industrial Classification*, e.g. steel making or sugar refining. Thus, the Census figures relate to distinct industries rather than companies with a wide range of activities.

The Census does not attempt to collect data about profits. There are problems in defining profit. And many establishments form part of large enterprises where profit cannot be easily calculated at establishment level. Instead, the Census provides figures of net output which, for most practical purposes, can be regarded as added value. Net output is calculated by first adjusting the sales revenue for the increase or decrease of stocks of finished goods and work-in-progress, to derive the gross output. Then the purchases of materials, components, fuel and certain other costs are deducted. Net output differs slightly from added value. It includes such costs as advertising, insurance, professional services, postage and telephones, rates, vehicle licence, plant hire and repairs. Also, from 1973 onwards, transport costs were included in net output. Another difference is that added value should include the employers' contributions to national insurance and pensions. Net output excludes such costs.

Fortunately, the difference between net output and added value is fairly small in most industries. According to the 1968 Census data, net output was £15,290 million. The various services listed above came to £1,484 million and the national insurance and pension contributions were £610 million. Thus the added value would have been £14,416 million, a difference of only 6 per cent from the net output figure. However, in a few industries certain costs such as advertising are much higher than average. So care must be taken in interpreting the data. Also, a few industries have high transport costs. So the figures for 1973 onwards must be treated with caution. Nevertheless, net output represents the value added to materials by the processes of production. There is no appreciable duplication involved in adding togather the net output of a number of industries.

From the Census data various ratios can be derived. Figure 2.3 shows the net output per head, an approximate measure of manpower productivity, for each of the 20 Orders. In each case, the total net output for the whole industry has been divided by the total number of people engaged in the industry, from shop-floor workers to

Figure 2.3 Net output per head - from Census of Production

Order	Industry	Net output per head, £			% of all mfg industries		
		1963	1968	1973	1963	1968	1973
III	Food, drink & tobacco	1686	2323	4441	123	118	126
IV	Petroleum products	2516	4375	7779	184	223	222
V	Chemical industries	2256	3273	6034	165	167	172
VI	Metal manufacture	1447	1958	3603	106	100	103
VII	Mechanical engineering	1370	2009	3325	100	102	95
VIII	Instrument engineering	1206	1749	2719	88	89	77
IX	Electrical engineering	1246	1825	3139	91	93	89
X	Shipbuilding	1057	1461	2518	77	74	72
XI	Vehicles	1454	2002	3205	106	102	91
XII	Other metal goods	1224	1719	2996	89	87	84
XIII	Textiles	1058	1588	2879	77	81	82
XIV	Leather and fur	1107	1512	2672	81	77	76
XV	Clothing and footwear	769	1088	2127	56	55	52
XVI	Bricks, pottery, glass, cement	1365	1958	4117	100	100	117
XVII	Timber, furniture	1147	1754	3814	84	89	109
XVIII	Paper, printing, publishing	1448	2049	3855	106	104	110
XIX	Other manufacturing industries	1266	1855	3327	93	94	95
III-XIX	All manufacturing industries	1361	1954	3497	100	100	100
II	Mining and quarrying	1190	1649	2914	87	84	83
XX	Construction	1093	1699	N/A	80	86	N/A
XXI	Gas, electricity, water	2417	3843	5741	177	196	164

Figure 2.4 Annual capital expenditure per head

Order	Industry	Annual capital expenditure per head, £			% of all mfg industries		
		1963	1968	1973	1963	1968	1973
III	Food, drink and tobacco	182	251	466	141	128	146
IV	Petroleum products	389	2255	1318	301	1156	414
V	Chemical industries	297	534	616	230	273	193
VI	Metal manufacturing	251	226	598	194	115	188
VII	Mechanical engineering	92	133	216	71	68	67
VIII	Instrument engineering	73	106	227	56	54	71
IX	Electrical engineering	87	123	201	67	63	63
X	Shipbuilding	79	133	250	61	68	78
XI	Vehicles	129	149	216	100	76	67
XII	Other metal goods	94	138	191	72	70	60
XIII	Textiles	94	188	342	72	96	107
XIV	Leather and fur	47	68	206	36	34	64
XV	Clothing and footwear	22	41	84	17	21	26
XVI	Bricks, pottery, glass, cement	170	296	492	131	151	154
XVII	Timber, furniture	67	111	304	51	56	95
XVIII	Paper, printing, publishing	133	165	270	103	84	84
XIX	Other manufacturing industries	119	230	396	92	117	124
III-XIX	All manufacturing industries	129	195	318	100	100	100
II	Mining and quarrying	150	214	934	116	109	293
XX	Construction	64	102	N/A	47	52	N/A
XXI	Gas, electricity, water	1613	2056	2411	1250	1054	758

managers, directors and working proprietors. Clearly, there are wide disparities between different industries. Comparison is facilitated by expressing each industry's figure as a percentage of the average for all manufacturing industries. Pride of place is taken by the petroleum products industry. It usually produced a net output per head more than double the average for all manufacturing industries. Second place in the table is taken by the gas, electricity and water industries; third place by the chemical industry. At the other extreme, bottom place is taken by the clothing and footwear industries. They have a net output per head only just over half of the average for all manu-facturing industries. The shipbuilding and marine engineering industry vies with textiles and with leather and fur for the place next to the bottom.

Some of the disparity between the industries with high net output per head and those with low figures is associated with differences in capital intensity. Oil refining and the chemical industries have considerable amounts of capital per employee. The clothing and textile industries tend to have small amounts of capital per employee. The Census does not collect data on the total capital employed. (If it did, there would be problems of definition and valuation.) Fortunately, it does collect data on annual capital expenditure. The annual capital expenditure per head can be used as a measure of the relative capital intensity of different industries. Those industries which are already capital-intensive tend to spend more per head on new capital expenditure than those industries which are already labour-intensive.

Figure 2.4 shows the annual expenditure per head for each industry. It can be seen that, despite some fluctuations, there is a reasonable consistency when each industry is compared with the average for all manufacturing industries. The gas, electricity and water industries invested 7 to 12 times the national average per head. The petroleum products industry invested 3 to 11 times the average. The chemical industry invested 2 to 3 times the average. At the other extreme, the clothing industry invested only a fifth of the average annual capital expenditure per head. The shipbuilding industry invested only three-quarters of the national average.

Thus the industries with the highest annual capital expenditure per head show the highest net output per head and vice versa. However, there are some anomalies. For example, the metal manufacturing industries invested more per head than the food, drink and tobacco industries, yet show lower net output per head. The timber and furniture industry invested less per head than the textiles, yet produced higher net output per head. Detailed analysis of the Census data for the 150 industries shows quite clearly that some industries with high capital expenditure per head have lower manpower productivity (expressed as net output per head) than other industries with much lower annual capital expenditure per head.

From Figure 2.3 it can be seen that all the industries showed an increase in the monetary value of the net output per head over the years. But the main reason for the increases was inflation, especially between 1968 and 1973. Nevertheless, there was some real growth over the years. If the figures of net output per head for all manufacturing industries are converted to 1970 prices throughout, the figures for 1963, 1968 and 1970 respectively are £1800, £2183 and £2631. Similarly, the annual capital expenditure per head for all manufacturing industries for the three years at 1970 prices were £165, £219 and £231.

More important than the absolute figures are the changes in ranking in some major industries. In terms of net output per head the vehicle industry was 6 per cent above the average in 1963. By 1973 it was 9 per cent below the average. Conversely, timber and furniture rose from 16 per cent below average in 1963 to 9 per cent above average in 1973. (The change in bricks, pottery, glass and cement can be accounted for mainly by the high transport costs included in the 1973 net output but not in 1968.) Detailed analysis of the Census data shows that the rise in net output per head was much faster in some of the 150 industries than others. The changes cannot be accounted for purely in terms of rates of capital expenditure. It seems that there is no guarantee that high rates of investment will increase manpower productivity. Nor does a modest rate of investment prevent increases in net output per head.

The relationship between capital expenditure and manpower productivity is intriguing. But an even more fascinating study is the relationship between manpower productivity and wage levels. The basic principle is that industries with high net output per head can afford to pay higher wages and salaries per head than industries in which net output per head is comparatively low. Of course, two or more industries with the same net output per head might have different rates of capital expenditure per head. Thus depreciation will differ. So the amounts left over from net output after depreciation will differ, leaving leaving less for wages and salaries in the industries where the rate of investment is higher. Another point is that the Census net output includes the cost of certain services. Some industries have a high proportion of such costs, e.g. advertising.

Figure 2.5 shows the average wage/salary per employee for the major Orders. The figures increase each year, partly because of inflation, though there was some real growth. The remarkable feature is that, when the figure for each industry is expressed as a percentage of the average for all manufacturing industries, the ratios are almost stable from year to year. It is almost as though an unseen

Figure 2.5 Average wage/salary per employee

Order	Industry	Average wage/salary per employee, £			% of all mfg industries		
		1963	1968	1973	1963	1968	1973
III	Food, drink and tobacco	655	904	1572	90	90	90
IV	Petroleum products	910	1212	2147	125	120	123
V	Chemical industries	834	1176	2008	115	117	115
VI	Metal manufacturing	827	1105	1966	114	110	113
VII	Mechanical engineering	792	1084	1836	109	108	106
VIII	Instrument engineering	716	985	1198	98	98	94
IX	Electrical engineering	712	954	1621	98	95	93
X	Shipbuilding	777	1091	1898	107	108	109
XI	Vehicles	873	1187	2104	120	118	121
XII	Other metal goods	683	941	1583	94	93	91
XIII	Textiles	568	814	1414	78	81	81
XIV	Leather and fur	624	830	1427	86	82	82
XV	Clothing	478	661	1070	65	65	61
XVI	Bricks, pottery, glass, cement	745	1024	1806	102	102	104
XVII	Timber, furniture	708	989	1794	97	98	103
XVIII	Paper, printing, publishing	796	1106	1964	109	110	113
XIX	Other manufacturing industries	672	944	1621	92	94	93
III-XIX	All manufacturing industries	725	1002	1732	100	100	100
II	Mining and quarrying	751	962	1671	103	96	96
XX	Construction	803	1119	N/A	110	111	N/A
XXI	Gas, electricity, water	834	1124	2048	115	112	118

Figure 2.6 Net output related to wages and salaries

Order	Industry	Wages and salaries as % of net output			Net output per £ of wages and salaries		
		1963	1968	1973	1963	1968	1973
III	Food, drink and tobacco	38.6	38.7	35.2	2.59	2.58	2.84
IV	Petroleum products	36.1	27.8	27.5	2.77	3.60	3.63
V	Chemical industries	36.8	35.8	33.2	2.72	2.80	3.02
VI	Metal manufacture	56.9	56.3	54.3	1.76	1.78	1.84
VII	Mechanical engineering	57.2	53.9	54.5	1.75	1.86	1.83
VIII	Instrument engineering	58.5	55.7	59.8	1.71	1.80	1.67
IX	Electrical engineering	56.9	52.1	51.4	1.76	1.92	1.94
X	Shipbuilding	73.0	74.3	74.9	1.37	1.35	1.33
XI	Vehicles	59.9	59.1	65.6	1.67	1.69	1.53
XII	Other metal goods	54.5	53.7	51.7	1.83	1.86	1.93
XIII	Textiles	53.3	50.1	48.8	1.88	1.97	2.05
XIV	Leather and fur	54.2	53.1	51.4	1.84	1.88	1.95
XV	Clothing and footwear	61.1	59.9	56.8	1.64	1.67	1.76
XVI	Bricks, pottery, glass, cement	54.0	51.8	43.4	1.85	1.93	2.30
XVII	Timber, furniture	59.2	54.4	45.5	1.69	1.84	2.20
XVIII	Paper, printing, publishing	53.9	53.1	50.0	1.85	1.88	2.00
XIX	Other manufacturing industries	52.3	50.3	48.1	1.91	1.99	2.08
III-XIX	All manufacturing industries	52.7	50.8	49.0	1.90	1.97	2.04
II	Mining and quarrying	63.1	58.3	57.2	1.59	1.72	1.75
XX	Construction	69.0	61.8	N/A	1.45	1.62	N/A
XXI	Gas, electricity and water	34.5	29.2	35.7	2.90	3.42	2.80

force determined that the wage and salary bill and net output rose hand-in-hand.

The wage/salary levels in Figure 2.5 can be compared with the levels of net output per head in Figure 2.3. In nine out of the ten Orders where net output per head was above average, the wage/salary level was also above average. The exception was food, drink and tobacco. However, in the ten Orders where net output per head is below average, two have wage/salary levels above average. These are shipbuilding and construction. It is noteworthy that these Orders employ mainly male workers, The food, drink and tobacco industries have high proportions of female employees. So do most of the other Orders with wage/salary levels below average.

Further analysis of the Census data reveals that over the period 1963 to 1973 there were 20 of the 150 industries where the net output per head was below average, yet the wage/salary level was above average. Conversely, there were only ten industries with net output per head above average yet with the wage/salary per employee below average. Thus high wages do not always go with high net output per head. As a broad generalisation, it is true. But there are sufficient exceptions to show that other factors help to determine wage and salary levels. Nevertheless, the best way to raise wage and salary levels is to increase the net output per head. The Census data shows that capital investment is a key factor. But it is not the only answer.

Another way of looking at the relationship between wages and net output is to express the total wage and salary bill as a percentage of the net output. The inverse of this figure, the net output per £ of wages and salaries expresses the same ratio in terms of the amount of net output generated from each £ of wage/salary. Figure 2.6 gives the data for the 20 Orders. The effect of the inclusion of transport costs in net output in 1973 is that, in most industries, the first ratio is reduced and the second ratio is increased. The effect is particularly noticeable in Order XVI covering bricks, pottery, glass and cement.

The ratios vary widely between the Orders. But in general, industries with high capital expenditure per head tend to show a high ratio of net output per £ of wages and salaries. The proportion of net output going to wages and salaries is higher in the labour-intensive industries where the capital expenditure per head is low. More remarkable is the fact that, within each industry, the ratios change very little from year to year (apart from the effect of transport costs in 1973). However, there are some noteworthy exceptions.

In the vehicle industries, the ratio of net output per £ of wages and salaries fell from 1.67 in 1963 to 1.53 in 1973. What this means is that, after paying out £1 of wage/salary in 1963 the industry had

67p left over to finance investment and cover other costs, excluding transport. By 1973 the industry had only 53p left over to cover investment and other costs, including transport. The vehicle sector covers aerospace, tractors, motor cycles and railway carriages and locomotives as well as motor vehicle manufacturing. If the latter industry is singled out, its index fell from 1.82 in 1963 to 1.41 in 1973. In terms of this index the motor vehicle industry was about 5 per cent below the average for all manufacturing industries in 1963. But by 1973 it was over 30 per cent below the average. Conversely, this index in the timber and furniture industries rose from 11 per cent below the average for all manufacturing industries in 1963 to 8 per cent above average in 1973.

Managers, trade unionists and others could learn some salutary lessons by comparing their own figures with the data in the *Census of Production*. How does your company compare with the national average in your sector? Are your wage and salary levels above or below the average in your industry? Are you investing too little compared with your competitors? Which industries offer the best opportunities for high manpower productivity and high wage and salary levels? The *Census of Production* is a goldmine of information for those who are prepared to dig.

ADDED VALUE

3

and business performance

3.1 Profit as a measure of performance

The traditional measure of business performance is profitability, the ratio of profit to capital employed. Some business men also use the ratio of profit to sales as an index of profitability. In fact the two indices are interrelated as shown in Figure 3.1.

The concept of profitability has some merits, especially for the investor. But it also has some serious defects. First, as a measure of performance, it can be very misleading. Second, in the modern climate of public opinion, it takes a somewhat narrow view. Third, it cannot be applied to non-profit-seeking organisations which nevertheless need to measure and improve their performance.

One of the problems with profit is the difficulty of definition. In theory, two companies could be identical in terms of the types of products, sales revenue, materials used, numbers employed, wage levels, capital employed, etc. Yet they could have different profit figures depending on the accounting practices adopted.

For example, the treatment of depreciation may vary. One company may write off its equipment over a period of 10 years. Another company may decide to depreciate over a 15-year period. The company with the lower depreciation figure will therefore show a higher profit after depreciation. Yet most financial pundits use the profit after depreciation as the basis for measuring return on capital. So, to compare two companies in terms of profitability, it is important to compare their depreciation policies too. Of course, the

Figure 3.1 Profit, sales and capital

	Profit margin on sales	x	Capital circulation	=	Return on capital employed
or	$\dfrac{Profit}{Sales}$	x	$\dfrac{Sales}{Capital}$	=	$\dfrac{Profit}{Capital}$

For example

	$\dfrac{£50,000}{£1,000,000}$	x	$\dfrac{£1,000,000}{£500,000}$	=	$\dfrac{£50,000}{£500,000}$
or	5%	x	Twice a year	=	10%

company with the higher depreciation policy will be showing a lower figure for capital employed. Nevertheless, the ratio of profit to capital could vary for no other reason than a difference in depreciation policy.

Similarly, two otherwise identical companies might have different policies for stock valuation. One firm might take a prudent view of writing down its stocks, thus reducing the declared profit. Another might take a more optimistic view, declaring a higher profit with not unreasonable stock values. A comparison of profitability in terms of return on capital would show different performances. Yet the reason for the difference is nothing to do with the quality of management of the resources. It arises purely from the way that an accounting procedure deals with the directors' opinions of stock values. Tax policy, too, can affect stock values and therefore distort profit figures.

Another factor affecting the profit figure is the treatment of development costs. Some companies engaged in activities with high development costs, e.g. aerospace, adopted a practice of writing off the huge development costs over several years. This practice shows a higher profit than if all the development costs are written off in the current year. Which profit figures should be used as the basis of measuring company performance?

Similar problems arise with the treatment of government grants. Some companies treat the grant as income and therefore part of the profit. Others deduct the grant from the capital expenditure so that only the balance sheet is affected, not the profit figure. Some enter the grant on the balance sheet and gradually transfer it to the profit and loss account. Thus any comparisons of profitability should take account of whether the company has received government grants and, if so, how the grants have affected the profit figure.

Yet another distortion of profit figures has arisen in recent years from the practice of financing expansion by means of loan stock rather than equity capital. If the rate of interest on loan stock is low in relation to the rate of profit on equity capital the company shows a better return on capital than if it is financed entirely by equity capital. Conversely, a high rate of interest on loan stock reduces the return on capital. Similar arguments apply for any form of loan. It might be more correct to express profitability in terms of profit plus interest charges as a ratio of shareholders' capital employed plus loans.

Finally, profit depends on the levels of wages and salaries. If one of two otherwise identical companies pays lower wages than the other it will show higher profits. Is it therefore more efficient in the use of resources? The investors might argue so. But from a broader perspective, the efficiency in the use of resources is identical. The two companies are equally efficient in generating wealth. They differ only in the way that the wealth is shared out between employees and investors. The difference is important. But it has nothing to do with efficiency in the use of resources.

Figure 3.2 gives some illustrations of differences in profitability that would arise from differences in depreciation policy, sources of finance and levels of wages. All the companies are identical in terms of gross output, purchases and added value. Company A has a policy

Figure 3.2 Comparisons of return on capital

Company (£'000s)	A	B	C	D
Sales, adjusted for stock change	1000	1000	1000	1000
Purchases, adjusted for stock change	400	400	400	400
ADDED VALUE	600	600	600	600
Wages, etc.	450	450	450	425
Depreciation	100	75	75	75
Interest on loans	–	–	25	25
Profit	50	75	50	75
Shareholders' capital	500	525	325	325
Loans	–	–	200	200
Return on capital, %	10	14.3	15.4	23

of higher depreciation than Company B so it shows a lower return on capital. Company C is financed partly by loan stock at an interest rate of 12.5 per cent. It therefore shows a higher return on equity capital than Company B. Company D has a lower wage level than Company C so it shows an even higher return on capital. This simple example could be extended to show the effect on profitability of the treatment of development costs, government grants, income from investments, etc.

A further problem in comparing returns on capital is that of asset valuation. Accountants beg to differ on questions of stock valuation and depreciation policy. Expert opinions on the value of land and buildings may vary widely. High rates of inflation have made a mockery of balance sheet values based on historical costs. Even without inflation it can be argued that the asset value depends on the profit record and potential rather than on the historical price or replacement cost. Finally, two companies with identical total assets could show different figures of capital employed if one has a higher proportion of external liabilities in the form of creditors, overdraft and other loans.

All these factors make it difficult to compare one company with another in terms of return on capital. Even comparisons within one company over periods of time may be distorted by some of the factors outlined above. Inflation accounting techniques can help to reduce the distortion of return on capital ratios. But profitability can be very misleading as an index of company performance.

Even if the technical problems of defining profit and capital could be overcome, there are emotional problems of using return on capital to measure company performance. Profit is seen by some people as unnecessary and evil. The word is associated with nineteenth century capitalism and the exploitation of workers. Explanation can be given of the need for profit as a reward for risk and as a source of funds for further investment. But some people have already closed their minds to such arguments. The social climate has changed with the declining power of individual capitalists and the rising power of the trade unions and government. A wider view of business performance is needed. It must take account not just of investors but of employees, customers, suppliers and governments.

Moreover, an increasing proportion of the population is now engaged in organisations which are not subject to the test of profit. So it is vital to develop measures of performance that can be applied to non-commercial activities. Thus profit is not just misunderstood. It is perhaps outmoded. New measures of performance must be developed to suit the modern world.

3.2   Added value as a measure of performance

If profitability is not reliable or acceptable as a measure of
performance, what is the alternative? Profitability relates a very
small part of the output, the profit, to only one of the factors of
production, the capital employed. What is needed is a broader
measure relating the total output to all the factors of production.
The appropriate word is productivity, the ratio of output to input.
Unfortunately, like profit, the word productivity is misused and
misunderstood. To some people productivity means working harder
for less pay. It is linked with redundancy and unemployment. But
productivity is the main determinant of standards of living throughout the world.

Productivity is not just a fancy word for production. It is the
fundamental ratio of the output to input of an organisation. It is
the relationship between, on the one hand, the output of goods and/
or services and, on the other hand, the input of resources used in
creating those goods and services. There is a common misconception
that productivity is concerned only with manpower — output per
man-hour. But there is no virtue in raising the productivity of manpower if the savings in labour costs are outweighed by increases in
the cost of materials or capital. Evaluations of productivity should
take account of all the resources used — materials, machinery, manpower, land, etc.

This theoretically simple concept is not easy to measure in
practice. Output can be expressed in terms of physical measures,
such as tonnes, metres or numerical quantities. But it is either
impossible or meaningless to add together quantities of different
kinds of output. For a small factory or department with only one
product, the physical measure of output is useful. But for the large
company with many diverse products, the total output cannot be
expressed in physical terms without the danger of over-simplification.

One way of overcoming this problem is to convert the physical
quantities of output into a common measure of time. For this
purpose, standard times based on work study can be used. If such
figures are not available, an alternative is to use estimated man-
hours or machine-hours, provided the estimates are consistent and
reliable. Thus the total output can be expressed in terms of standard
hours or estimated hours. However, if the standard time or estimated
time for a product or operation should change, through using
different methods or materials, the same physical output may give
a different total output in terms of time. Another shortcoming is
that a measure based on time ignores the value to the customer. An
article may take several hours to make. But if it cannot be sold, it
should not really be counted as part of the output.

Physical measures and time measures of output are useful at plant or departmental level. But they are not adequate for measuring the output of a company. The most useful numerator in an index of productivity is the monetary value of the output. Of course, monetary values may be distorted by inflation. But this problem can be tackled by using suitable index numbers. The great merit of monetary measures is that they reflect the satisfaction of the customer. They also measure the collective value of all the factors of production.

Most industrial and commercial enterprises express their output in terms of sales turnover. Many managers find it useful to look at such indices as the sales per employee, sales per £ of wages, sales per £ of capital employed, etc. Unfortunately, sales turnover is not necessarily a reliable monetary measure of output. In any particular period, the sales turnover may be less than the actual output, so stocks of finished goods will rise. Alternatively, stocks may fall if sales turnover exceeds the level of output. This drawback can be overcome by making appropriate adjustments for stock changes in order to derive the gross output.

Even then, comparisons in terms of gross output can be misleading, whether between companies or within one company. Both sales turnover and gross output include the cost of all the materials, bought-in parts, fuel, transport and other services purchased by the enterprise. They include the wealth created by other enterprises. The distinction is shown quite clearly in Figure 3.3.

Figure 3.3 Sales turnover and added value

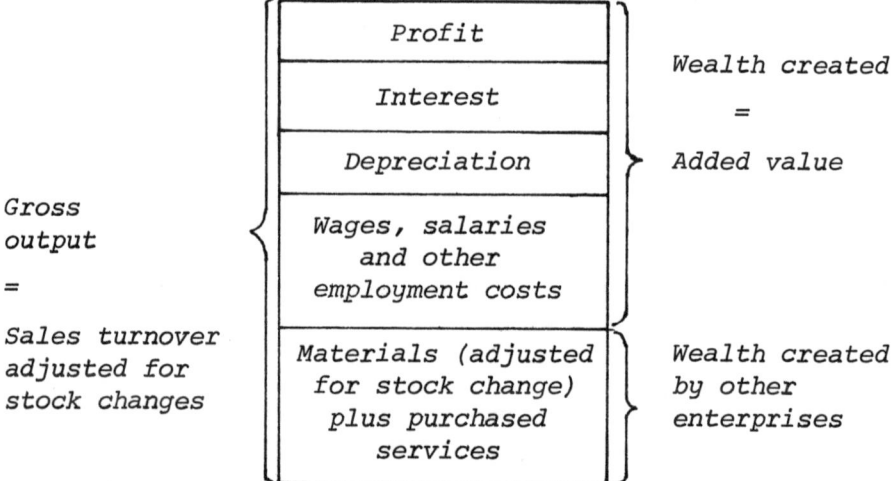

38

The importance of added value can be seen from Figure 3.4 showing data based on an actual company. The two columns compare consecutive years. The gross output rose by 10 per cent (before the days of high inflation). But the whole of the extra income was swallowed up by higher costs of materials and purchased services. Thus the added value was unchanged. The managing director had proudly boasted about the increase in output. He soon came to realise that, in reality, the company was no better off. Indeed, it was worse off. It had financed the extra stocks and debtors from the higher output but had nothing to show for it.

Figure 3.4 Sales turnover - an unreliable measure

£'000s	Year 1	Year 2
Sales, adjusted for stock change	1000	1100
Purchases, adjusted for stock change	600	700
ADDED VALUE	400	400

When it comes to comparing one company with another, the defects of sales turnover are even more apparent. A manufacturing business with the same sales turnover as a retail business usually has a much smaller proportion of purchases. The ratio of purchases to sales also varies between manufacturing industries, from 80 down to 30 per cent. The contrast is illustrated in Figure 3.5. Thus sales turnover, even after adjustment for stock changes, has serious defects as a measure of output. A more useful measure than sales turnover is added value, the difference between the value of the goods produced and the cost of the materials used and the services purchased. Added value discounts the effect of differences in material

Figure 3.5 Retailing and manufacturing

£'000s	Retailing	Manufacturing	
		A	B
Sales, adjusted for stock change	1000	1000	1000
Purchases, adjusted for stock change	750	600	400
ADDED VALUE	250	400	600

costs. More important, it represents the value added to the materials by the processes of production. It measures the wealth created by an enterprise, whereas sales turnover includes the wealth created by other enterprises. Added value is the sum available to cover all employment costs, depreciation, interest charges and profit.

The added value concept overcomes the problems of using sales turnover or profit as a measure of output. But in order to establish a measure of performance, the output must be divided by the inputs. The main inputs are materials, manpower and capital. The use of added value for measuring output discounts the cost of materials. So the main inputs are manpower and capital. The index of productivity can then be expressed in terms of the ratio below:

$$\text{Productivity} = \frac{\text{Output}}{\text{Input}} = \frac{\text{Added value}}{\text{Manpower} + \text{Capital}}$$

Unfortunately, there is no easy way of adding together manpower and capital. The problem has baffled economists, accountants and managers for many years. Various ideas have been put forward for converting the value of capital into manpower equivalent. Alternatively, some people have tried to convert the wage bill into a capital sum. For a variety of reasons, no answer has proved to be universally acceptable. Indeed, the search can be likened to the quest of the alchemists to transmute lead into gold.

Instead of attempting to achieve the difficult task of adding together manpower and capital, performance can be compared in terms of the trends of manpower productivity and capital productivity over periods of time. What matters is not so much the ratios in one particular period, but the trends. If added value per employee is rising and, at the same time, the added value per unit of capital is also rising, the rewards to both employees and investors can increase. If both ratios are falling, the reward to one or both

Figure 3.6 Manpower productivity and capital productivity

Manpower productivity (added value per employee)	Capital productivity (added value per unit of capital)		
	Rising	Static	Falling
Rising	Excellent	Good	Poor
Static	Good	Static	Bad
Falling	Poor	Bad	Very bad

must suffer. One index rising and the other static is a better situation than both static or one falling whilst the other is static. There are nine possible combinations of rising, static or falling productivity of manpower and capital. These are illustrated in Figure 3.6. The nine basic combinations could be extended to 25 to allow for the rate of rise or fall in productivity. A company showing a rapid rise in both manpower and capital productivity is better than one where both ratios are rising slowly. Conversely a company where both ratios are falling rapidly is worse than one in which both ratios are falling slowly. It is difficult to compare a company where manpower productivity is rising and capital productivity is falling with a company where manpower productivity is falling and capital productivity is rising. Nevertheless, there is little doubt that both companies would be better off if they could arrest and reverse the ratio that is falling.

To use the concept of trends, manpower productivity can be measured in terms of added value per employee. It may be necessary to make adjustments for inflation if the rate of inflation is high compared with the increase in real output per employee. It is more difficult to find a satisfactory measure of capital. The conventional capital employed figure used for profitability has the limitations discussed earlier. It may be better to use the figure of total assets rather than just shareholders' capital. Alternatively, it may be appropriate to concentrate on the fixed assets rather than include the current assets. But whatever the measure used, it is important to adjust for inflation, using one of the recognised accounting procedures.

The relationship between added value, sales and capital is shown in Figure 3.7. The formula is similar to that shown in Figure 3.1

Figure 3.7 Added value, sales and capital

$\dfrac{\text{Added value}}{\text{per £ of sales}}$	x	$\dfrac{\text{Capital}}{\text{circulation}}$	=	$\dfrac{\text{Added value}}{\text{per £ of capital}}$
or $\dfrac{\text{Added value}}{\text{Sales}}$	x	$\dfrac{\text{Sales}}{\text{Capital}}$	=	$\dfrac{\text{Added value}}{\text{Capital}}$

For example

$\dfrac{£400,000}{£1,000,000}$	x	$\dfrac{£1,000,000}{£500,000}$	=	$\dfrac{£400,000}{£500,000}$
or 40%	x	Twice a year	=	80%

but added value has replaced profit as the numerator. The ratio of added value to sales and the ratio of added value to capital are key indices in a business. In order to improve the ratio of added value to capital, the ratio of added value to sales must be increased and/or the capital circulation must be improved. The ratio of added value to sales can be improved by making better use of materials, by improving the product mix or by raising selling prices. The capital circulation can be increased by producing more output from existing equipment and buildings.

However, care must be taken in comparing companies in terms of their ratio of added value to sales. A business that makes a complex product from basic raw material will have a higher ratio of added value to sales than a company that simply assembles components made by other firms. There is no merit in a high or low ratio as such. But there is merit in raising the ratio in a given type of business.

Another way of using added value for performance measurement is to relate the added value per employee to the capital per employee. In theory, a business with high capital per employee should generate more added value per head than one with low capital per employee. In practice, different firms with the same amount of capital per employee show marked differences in added value per employee.

However, such comparisons must be treated with caution. It is vital to compare like with like. A business that rents its land and buildings and hires its plant will have low figures of capital per employee. Even in businesses that do not rent or hire their fixed assets, depreciation will vary from one company to another. Nevertheless, if proper care is taken, valid comparisons can be made. One thing is certain. Ratios that relate added value to the number of employees and to capital employed are less distorted by inflation and accounting conventions than are ratios of profit to capital.

3.3 Business performance — a case history

To illustrate the use of added value ratios in practice, a case history based on an actual company is now described. The company is Imperial Chemical Industries Limited, one of Britain's largest companies. The main reason for choosing ICI is simple. It is one of the few companies that have revealed enough information in their published accounts to permit the calculation of added value. Under the Companies Acts of 1948 and 1967, companies are not obliged to disclose their expenditure on materials and purchased services. Very few companies have revealed such figures.

ICI is one of the exceptional companies that have gone beyond the requirements of the law. In theory, added value can be calculated by adding together employment costs, depreciation, interest charges and profit. In practice, there are the complications of government grants, investment income, royalties, profit from associated companies, etc. Another problem is that companies are obliged to disclose the remuneration of UK employees but not of any overseas employees.

ICI was one of the first companies to publish an added value statement with their annual report. It first appeared in their 1975 annual report, quoting figures for 1974 and 1975. Using this format, similar data for earlier years can be extracted from the published annual reports. Figure 3.8 sets out the data for the 8 years from 1968 to 1975. The top section shows the various sources of income. The major sources are the sales of UK and overseas companies within the ICI group. But there are also some minor items that make up about 3 per cent of the total income. The next line shows the cost of materials and purchased services. The difference between total income and the cost of materials and purchased services give the added value.

The remaining lines in the table show the disposal of the added value in terms of four categories. The major slice went to the employees in the form of wages, salaries, profit-sharing bonus and pension fund contributions. (The latter item includes redundancy pay in recent years.) The next slice was for the government in the form of corporation tax less investment grants and regional development grants. A third slice went to the providers of capital. The figures are broken down into interest on loans, dividends to shareholders and payments to minority shareholders in subsidiaries. The final slice was reinvested in the business in the form of depreciation and retained profit. (The government's slice included deferred tax which was also retained in the business.)

It should be noted that the added value statement published by ICI in 1976 uses a slightly different format and definitions. The main change is the inclusion of employer's national insurance contributions in the total pay. This adds about 6 per cent to the original figure. Unfortunately, the new figures cannot be compared with those published for the earlier years. However, the change does not invalidate the use of the earlier figures as a basis for commenting on business performance.

The monetary figures for sales, costs, added value, wages, etc., increase year by year. Part of the explanation is inflation but there is also some growth in real terms. It is difficult to convert the monetary figures into constant prices because ICI is a world-wide company. However, there is little doubt that over the 8-year period, the total output increased in real terms. But in some years the monetary increase may distort the extent of the increase in real terms, or even disguise a decrease.

Disposal of added value

To employees:								
Wages and salaries	248	277	318	354	417	474	565	686
Pension fund contributions*	17	18	21	30	37	57	76	82
Profit-sharing bonus	10	10	8	7	8	17	25	17
TOTAL	275	305	347	391	462	548	666	785
To government:								
Corporation tax	70	74	59	51	57	145	220	151
Tax on dividends	25	26	27	25	26	–	–	–
Less grants	11	14	17	19	24	27	33	43
NET TOTAL	84	86	69	57	59	118	187	108
To providers of capital:								
Interest to loan stockholders	30	32	35	43	52	59	64	72
Dividends to shareholders	35	37	38	40	41	50	54	59
Minority shareholders in subsidiaries	8	11	10	13	17	25	19	24
TOTAL	73	80	83	96	110	134	137	155
Re-invested in the business:								
Depreciation	105	107	114	129	148	157	169	181
Profit retained	26	33	27	20	24	134	189	132
TOTAL	131	140	141	149	172	291	358	313
= ADDED VALUE	563	611	640	693	803	1091	1348	1361

*Including redundancy payments in later years

Figure 3.8 ICI added value statement

Year ended December 31	1968 £m	1969 £m	1970 £m	1971 £m	1972 £m	1973 £m	1974 £m	1975 £m
Sources of income								
Sales – United Kingdom	616	652	693	704	776	942	1199	1311
Sales – overseas	621	703	769	820	918	1224	1756	1788
TOTAL SALES	1237	1355	1462	1524	1694	2166	2955	3099
Royalties and other trading income	19	16	18	13	19	19	22	30
Investment income	17	19	12	13	13	24	44	39
Profit from associated companies	–	–	16	22	18	34	43	24
TOTAL INCOME	1273	1390	1508	1572	1744	2243	3064	3192
Less: Cost of materials and purchased services	710	779	868	879	941	1152	1716	1831
= ADDED VALUE	563	611	640	693	803	1091	1348	1361

Figure 3.9   ICI added value statement in percentages

Year ending 31 December	1968 %	1969 %	1970 %	1971 %	1972 %	1973 %	1974 %	1975 %
Sources of income								
Sales – United Kingdom	48.4	46.9	45.9	44.8	44.5	42.0	39.1	41.1
Sales – overseas	48.8	50.6	51.0	52.2	52.6	54.6	57.3	56.0
TOTAL SALES	97.2	97.5	96.9	97.0	97.1	96.6	96.4	97.1
Royalties and other trading income	1.5	1.1	1.2	0.8	1.1	0.8	0.7	0.9
Investment income	1.3	1.4	0.8	0.8	0.8	1.1	1.5	1.2
Profits from associated companies	–	–	1.1	1.4	1.0	1.5	1.4	0.8
TOTAL INCOME	100.0	100.0	100.0	100.0	100.0	100.0	100.0	100.0
Less: Cost of materials and purchased services	55.8	56.0	57.6	55.9	54.0	51.4	56.0	57.4
= ADDED VALUE	44.2	44.0	42.4	44.1	46.0	48.6	44.0	42.6

Disposal of added value

To employees:								
Wages and salaries	44.1	45.4	49.7	51.1	51.9	43.4	41.9	50.4
Pension fund contributions	3.0	2.9	3.3	4.3	4.6	5.2	5.6	6.0
Profit-sharing bonus	1.8	1.6	1.2	1.0	1.0	1.6	1.9	1.3
TOTAL	48.9	49.9	54.2	56.4	57.5	50.2	49.4	57.7
To government:								
Corporation tax	12.4	12.1	9.2	7.4	7.1	13.3	16.3	11.1
Tax on dividends	4.4	4.3	4.3	3.6	3.3	-	-	-
Less grants	1.9	2.3	2.7	2.8	3.0	2.5	2.4	3.2
NET TOTAL	14.9	14.1	10.8	8.2	7.4	10.8	13.9	7.9
To providers of capital:								
Interest to loan stockholders	5.3	5.2	5.5	6.2	6.5	5.4	4.8	5.3
Dividends to shareholders	6.2	6.1	5.9	5.8	5.1	4.6	4.0	4.3
Minority shareholders in subsidiaries	1.4	1.8	1.6	1.9	2.1	2.3	1.4	1.8
TOTAL	12.9	13.1	13.0	13.9	13.7	12.3	10.2	11.4
Re-invested in the business:								
Depreciation	18.7	17.5	17.8	18.6	18.4	14.4	12.5	13.3
Profit retained	4.6	5.4	4.2	2.9	3.0	12.3	14.0	9.7
TOTAL	23.3	22.9	22.0	21.5	21.4	26.7	26.5	23.0
= ADDED VALUE	100.0	100.0	100.0	100.0	100.0	100.0	100.0	100.0

Figure 3.10 ICI: manpower productivity, wages and capital productivity

Year ending 31 December		1968	1969	1970	1971	1972	1973	1974	1975
Added value	£m	563	611	640	693	803	1091	1348	1361
Number of employees:									
United Kingdom	000s	139	145	142	137	135	132	132	129
Overseas	000s	48	52	52	53	64	67	69	66
TOTAL	000s	187	197	194	190	199	199	201	195
ADDED VALUE PER EMPLOYEE	£	3010	3101	3299	3647	4035	5482	6706	6979
Wages and salaries:									
United Kingdom	£m	188	214	241	262	281	313	374	459
Overseas	£m	60	63	79	92	136	161	191	227
TOTAL	£m	248	277	318	354	417	474	565	686
Profit-sharing bonus	£m	10	10	8	7	8	17	25	17
Pension fund contributions	£m	17	18	21	30	37	57	76	82
TOTAL EMPLOYMENT COSTS	£m	275	305	347	391	464	548	666	785
ADDED VALUE PER £ OF EMPLOYMENT COSTS		2.05	2.00	1.84	1.77	1.73	1.99	2.02	1.73

Average wage/salary:									
United Kingdom	£	1352	1475	1697	1912	2081	2371	2833	3558
Overseas	£	1250	1211	1519	1736	2125	2402	2768	3439
TOTAL	£	1326	1406	1639	1863	2095	2382	2811	3518
Fixed assets	£m	972	1016	1076	1157	1160	1185	1233	1425
Total assets employed	£m	1487	1557	1675	1774	1869	2143	2412	2748
Added value per £ of fixed assets		0.58	0.60	0.59	0.60	0.69	0.92	1.09	0.96
Added value per £ of total assets		0.38	0.39	0.38	0.39	0.43	0.51	0.56	0.50
Inflation-adjusted fixed assets*							1955	2310	2281
Inflation-adjusted total assets*							3158	3889	3732
Added value per £ of fixed assets							0.56	0.58	0.60
Added value per £ of total assets							0.35	0.35	0.36

*Adjusted by current purchasing power method

Figure 3.11 ICI: profitability ratios

Year ending 31 December	1968	1969	1970	1971	1972	1973	1974	1975
A Trading profit after depreciation	175	190	159	145	169	329	457	353
+ Investment income	18	19	12	13	14	24	44	39
− Interest payable	30	32	35	43	52	59	64	72
− Employees' bonuses	10	10	8	7	8	17	25	17
+ Profit from associated companies	−	−	16	22	18	34	43	24
B Profit before tax and grants	153	167	144	130	141	311	455	327
− Taxation	70	74	59	51	57	145	220	151
+ Grants	11	14	17	19	24	27	33	43
− Extraordinary items	−	−	−	−	−	15	6	4
C Profit after tax and grants, etc.	94	107	102	98	108	208	262	215
D Ordinary capital of ICI	446	448	469	474	478	480	486	494
Group reserves (retained profit)	372	413	482	449	469	644	845	993
ICI shareholders' interest	818	861	951	923	947	1124	1331	1487
Minority interests and preference capital	140	122	125	123	133	165	187	203

Investment grants	48	67	80	91	81	67	53	36
Deferred taxation	107	95	77	62	91	115	174	212
Loans	374	412	442	575	617	672	667	810
E Total funds invested	1487	1557	1675	1774	1869	2143	2412	2748
F Total funds less loans	1113	1145	1233	1199	1252	1471	1745	1938
A/E Trading profit / Total funds	11.7	12.2	9.5	8.2	9.0	15.4	18.9	12.8
B/D Profit before tax / Shareholders' interest	18.7	19.4	15.1	14.1	14.9	27.7	34.2	22.0
B/F Profit before tax / Total funds less loans	13.7	14.5	11.6	10.8	11.3	21.1	26.1	16.9
C/D Profit after tax / Shareholders' interest	11.5	12.4	10.7	10.6	11.4	18.5	19.7	14.5
C/F Profit after tax / Total funds less loans	8.4	9.3	8.3	8.2	8.6	14.1	15.0	11.1

Another way of looking at the figures is to convert the monetary values into percentages of sales and added value. Figure 3.9 gives the results. It shows that added value as a percentage of total income varied from 42.4 to 48.6 per cent. The variation from the average of 44.5 per cent was quite small. What this means is that ICI managed to keep its selling prices broadly in line with the rising costs of raw materials and purchased services. The ratio of added value to total income was highest in the boom year of 1973. The ratio was lowest in the recession years of 1970 and 1975.

Figure 3.9 also shows that the total employment costs, as a percentage of added value, varied from 48.9 to 57.7 per cent. It is noticeable that in the years when this percentage was high, the government received a lower share. In the years when employment costs were a low percentage of added value, the tax proportion rose. In later years, the retained profit also increased.

There was a noticeable trend between 1968 and 1972. Employment costs, as a percentage of added value, rose from 48.9 to 57.5 per cent. The increase of 8.6 per cent in four years was quite marked. If this trend had continued, the ratio in 1975 would have been 64 per cent. By 1985 it would have been 85 per cent. By 1991 it would have reached 100 per cent. At some point ICI would have become unprofitable. Fortunately, in the boom year of 1973, the added value rose faster than wages and salaries.

There are no hard and fast rules to say what the ratio of employment costs to added value should be. But clearly, if the ratio had gone on increasing, ICI would have made losses rather than profits. Moreover, it would have been unable to pay a dividend and finance further investment. In principle, if the ratio can be maintained at a certain level the company can thrive and prosper to the benefit of employees, shareholders and customers.

But how did ICI fare in terms of manpower productivity and capital productivity? Some answers can be found from Figure 3.10. This shows, first, the number of employees in the UK and overseas. The total number of employees, both UK and overseas, is then divided into the total added value to show the added value per employee. The figure increases every year. By 1975 it was more than double the 1968 figure. Again, the figures are affected by inflation. Nevertheless, there is little doubt that, over the eight years, productivity in terms of added value per employee rose not just in monetary terms but also in real terms.

It is not possible to calculate the added value per UK employee and per overseas employee because the annual report does not show separate figures for purchases for UK and overseas. However, the sales turnover per overseas employee has been 2½ to 3 times the figure of sales per UK employee. But the mix of products was different. Even so, it seems unlikely that the added value per overseas employee was lower than the figure of added value per UK employee.

So, in ICI over this period, manpower productivity, measured in terms of added value per head, was rising. Perhaps it was not rising as fast as it could or should have done. Perhaps the levels of added value per head were below those of other international companies in similar fields. But at least the trend was rising, not falling.

At this state in the analysis, it is useful to look at the added value generated per £ of employment costs. Figure 3.10 shows the total remuneration of the UK and overseas employees, the profit-sharing bonus and the pension fund contributions. The sum total of these items is divided into the added value to show the added value per £ of employment costs. This index, of course, is the inverse of the ratio shown in Figure 3.8, expressing the employment costs as a percentage of added value. What the figures mean is that in 1968, after paying out £1 of wages, salaries and other employment costs, ICI had another £1.05 left over to cover depreciation, interest charges and profit before tax. In 1973 and 1975, they had only 73p left over for the same purposes. Bearing in mind that one of the effects of inflation is to understate the depreciation, the figures in 1973 and 1975 are even further below those of 1968 than appears at first sight.

The next three lines in Figure 3.10 show the average remuneration per employee, both UK and overseas. These figures exclude the profit-sharing bonus and pension fund contributions. It is interesting to note that in 1968, the average remuneration per UK employee was higher than that for overseas employees. This state of affairs continued until 1972 when the overseas employees started to earn more per head than the UK employees. Then in 1975 the UK employees once more received the higher earnings. However, the latter figures may be distorted by inflation which was higher in the UK than overseas.

What happened to capital productivity in ICI over this period? Figure 3.10 shows some figures, taken from the balance sheets, expressing the value of fixed assets, after depreciation, and the total assets employed. The latter figure is the total fixed assets, after depreciation, plus goodwill, trade investments, interests in subsidiaries and also plus the *net* current assets, i.e. total current assets less current liabilities. This sum also represents the funds invested by the shareholders and loan stockholders plus investment grants and deferred taxation.

These two sets of figures are a measure of the capital employed by ICI. Other measures of capital could be derived from the balance sheet, such as fixed assets before depreciation or grand total assets, i.e. fixed plus all current assets. The two shown in the table were chosen for simplicity in explanation.

These figures can be divided into the added value for the corresponding year to give two indices of capital productivity, namely added value per £ of fixed assets and added value per £ of total assets employed. It can be seen that these indices increased almost every year, especially in recent years. However, the figures are distorted by inflation because the balance sheet figures are based on historical asset values whereas the added value is based on current figures. Fortunately, ICI, again leading the way in the presentation of information in annual reports, issued some figures for balance sheet values adjusted for inflation. They used the current purchasing power method which has since found less favour in official circles than the current cost accounting method. Nevertheless, any adjustment is better than none.

It can be seen that, after adjustment for inflation, the indices of added value per £ of fixed assets, and per £ of total assets, in 1973, 74 and 75 are similar to the figures for 1968-71. Thus it can be argued that capital productivity in ICI was more or less static. It was not rising substantially, as appears before adjustment for inflation. But it was not falling.

To sum up, over the 8-year period, manpower productivity, expressed as added value per head, was rising, albeit slowly and erratically. Capital productivity expressed as added value per £ of fixed or total assets was static in real terms. So by the criteria laid down in Figure 3.6, ICI's performance was good but not excellent. ICI was improving the utilisation of its resources of manpower. At the same time, its utilisation of capital was not declining.

But that is not the end of the story. The question is whether this method of judging company performance is a better guide than the traditional measure of profitability, i.e. return on capital.

Figure 3.11 sets out the data from the company accounts. The top section shows various profit figures. It starts with the trading profit after depreciation. To this figure is added investment income and profit from associated companies. From this figure is deducted interest payable and the employees' profit-sharing bonus. The result is net profit before taxation and grants. Then the taxation is deducted and grants are added. In the last three years there are also some adjustments for extraordinary items, mainly the effect of changes in exchange rates.

The next section of Figure 3.11 shows the capital employed. It starts with the shareholders' stake, including retained profit. To this is added minority interests, preference capital, investment grants, deferred taxation and loans. The result is the total funds invested. This corresponds with the figures of total assets shown in Figure 3.10. The last line shows total funds less loans. This is the same as the sum total of shareholders' capital plus minority interests, preference capital, investment grants and deferred taxation.

The bottom section of Figure 3.11 shows various ratios of profit to capital. First comes the ratio of trading profit to total funds. This represents the performance from trading activities in relation to funds invested. The next ratio shows the net profit before taxation and grants related to the shareholders' capital. However, the latter figure

Figure 3.12 ICI added value and wages

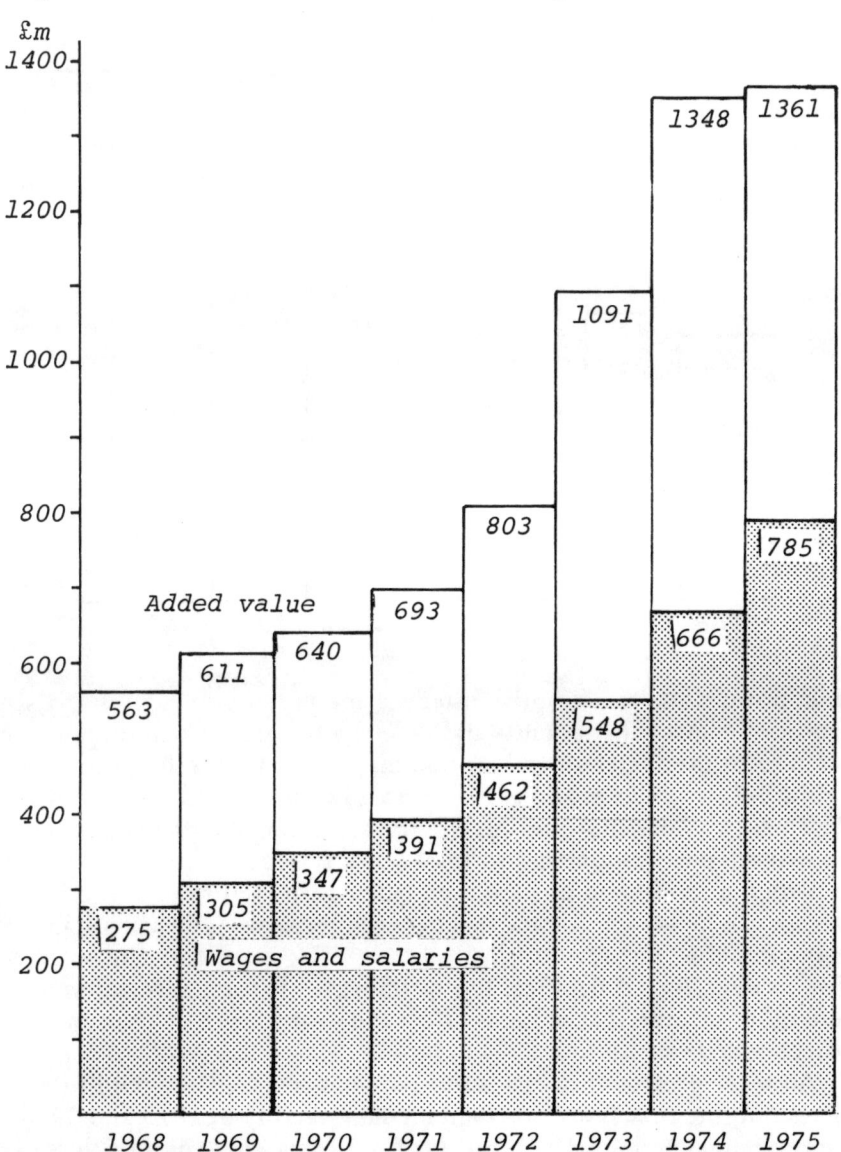

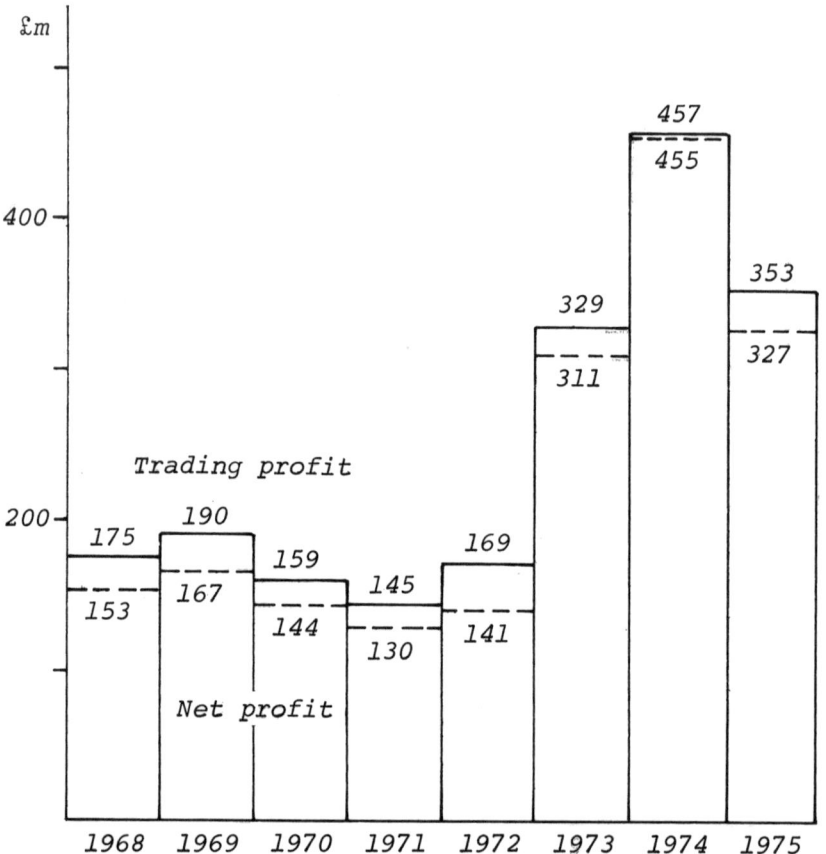

Figure 3.13 ICI trading profit and net profit

does not include the minority interests and preference capital. It also ignores the investment grants and deferred taxation which, in practice, are part of the shareholders' capital. So the third ratio shows the net profit before taxation and grants related to the total funds less loans. The last two lines show net profit, after taxation and grants, related to shareholders' capital and to total funds less loans.

Which of these figures should be taken as the return on capital? A case can be argued for and against each measure. Other profitability ratios could be put forward instead. One thing is certain. The profit figures are affected significantly by adjustments for grants, interest and other income. These items affect the profit far more than they affect added value.

All these profitability ratios show similar trends. They tend to decline in the first few years then increase sharply in 1973 and 1974 before falling again in 1975. But profit is a very small part of the

Figure 3.14 ICI added value per head and added value per £ of assets

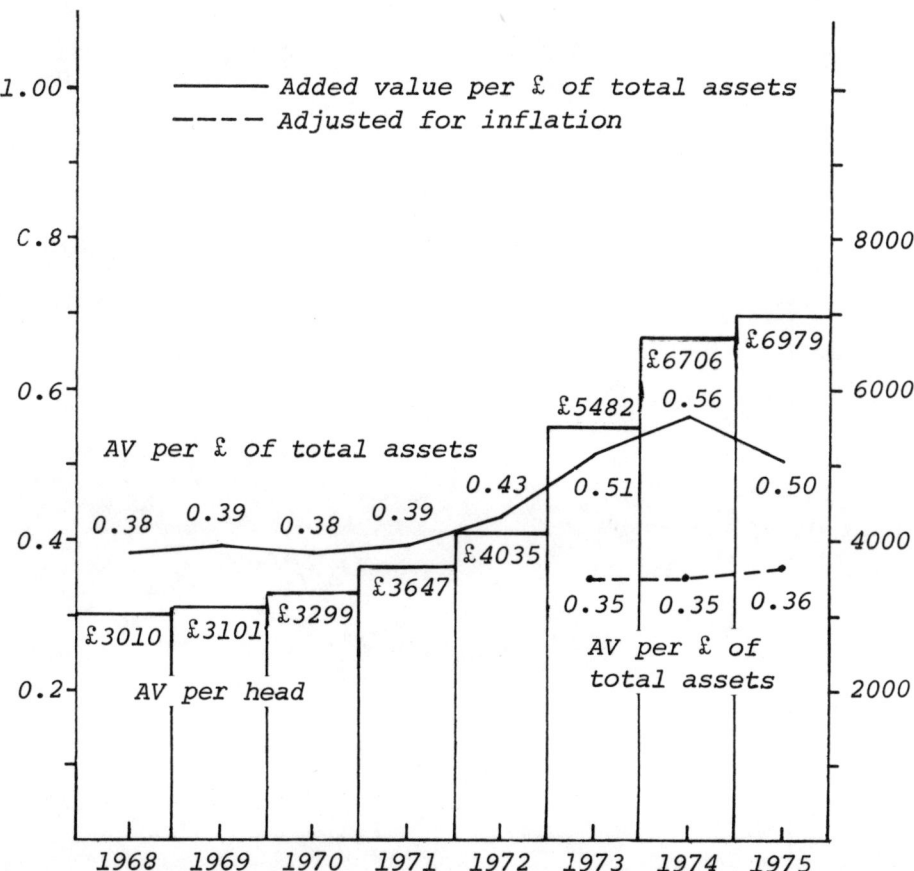

wealth created. So it tends to vary more widely than added value. The difference is illustrated in Figures 3.12-3.15.

Figure 3.12 shows the total added value and the total wages, salaries and other employment costs. The diagram reveals an upward trend every year. Figure 3.13 shows the trading profit and the net profit. The changes from year to year are quite erratic. Figure 3.14 shows added value per head and added value per £ of total assets. The first index moves steadily upwards, the second is fairly stable especially after adjustment for inflation. Figure 3.15 illustrates two indices of return on capital, both showing marked changes from year to year.

The real point about profitability indices is that they confuse two different issues. One factor is efficiency in the use of resources of manpower and capital. The second factor is how the wealth created

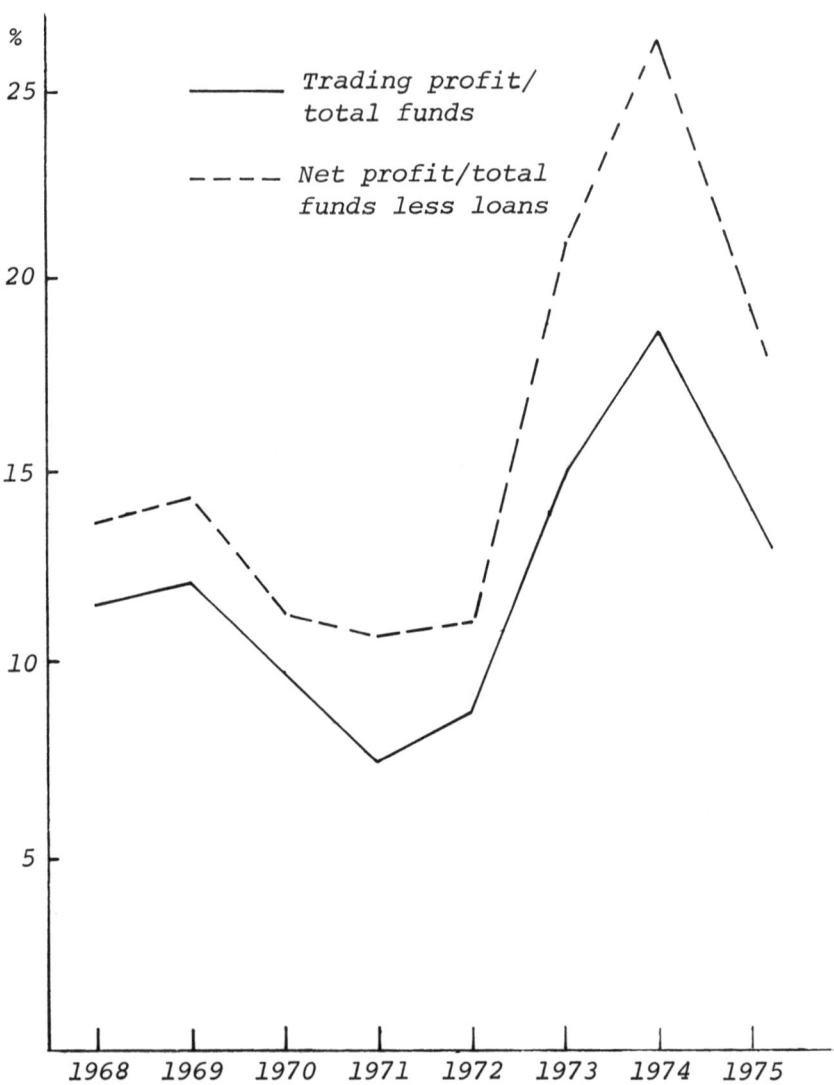

Figure 3.15 ICI return on capital

Trading profit/ total funds

Net profit/total funds less loans

is shared out. Profitability mixes these two issues. A company could be making good use of manpower and capital but showing low profits because a high proportion of the wealth created is going to the employees. Another company could be less efficient in the use of manpower and capital yet show high profits because only a low proportion of the wealth created is going to the employees.

Judging by the profitability measure, ICI's performance was declining between 1969 and 1971 then it rose rapidly until 1974 before falling in 1975.

Judged by the added value ratios, it was improving manpower utilisation almost every year and its utilisation of capital was not falling. In ICI between 1968 and 1972 the employees were receiving an increasing share of the wealth created. This does not mean that the company was less efficient in the use of resources. Yet the profitability measure confuses the two issues of efficiency and wealth sharing.

Of course, to survive in the long run without subsidy a business must not only make efficient use of its resources. It must also share out the wealth created in such a way that it can replace its assets as they wear out and also finance any expansion. Over the years, ICI has succeeded on both issues. Its profitability has fluctuated, especially with high inflation. But its productivity has improved and it has maintained a reasonable balance in sharing out the wealth created.

However, this method of using added value to measure company performance only compares the company with itself over a period of years. If ICI is compared with other international chemical companies in terms of added value per head and added value per £ of capital, it is nearer the bottom of the league table than the top. But ICI is not just in chemicals so such comparisons can be challenged. Nevertheless, it seems that ICI, like many other British companies, needs to increase its added value per head to match its international competitors. Also like other British firms, ICI's remuneration per employee is below the level of its competitors. If ICI could increase its manpower productivity and its capital productivity, it could pay higher wages and bigger dividends.

Other methods of using added value for measuring performance are discussed in Section 8.5.

4 ADDED VALUE

and communication

4.1 What business is about

What is the purpose of a business? Is it simply to make a profit for
the benefit of the shareholders? Or does a business have a wider
purpose? If so, what is business about?

To answer these questions we must first look at the history of
business. Before the industrial revolution most businesses were very
small. Many of them were family businesses in the strict sense. The
people working in the business were members of the same family.
There were traditional trades and crafts. The farmer, the builder,
the carpenter, the weaver and the tailor were the basic producers.
They sold their wares to one another and to the merchants, traders
and shopkeepers who sold the products to other customers. Most
of the businesses were labour-intensive. They needed very little
capital per person.

The industrial revolution saw the start of the factory system.
The family business expanded by taking on workers to operate the
new machines. These produced more and better products than the
traditional hand methods. Thus the business involved two distinct
groups of people. There were the capitalists who owned the factories
and machines. And there were the workers who sold their labour
by the hour or on piecework. As businesses grew, more capital
was needed. Some of this was provided by selling shares to people
outside the business. They invested their capital where they thought
they would get the best return compatible with the risk. Nowadays,

there are giant companies with shareholders who have never been inside the business. The link between ownership in the legal sense and involvement in the management sense has been broken.

In the small family business the word profit has a different meaning from the definition used in a large modern business. In both cases, profit is the difference between income and expenditure, the excess of returns over outlay. But when a small shopkeeper talks about making a profit he means the gap between what he receives from his customers and what he pays to his suppliers. This is his gross profit. From this he pays his rent, rates, electricity and telephone bill, etc., leaving him with a net profit. After paying income tax on his so-called profit, the shopkeeper is left with his net income to pay himself and his wife. He can choose either to spend all his income or to invest some of it in the business. But his only source of income is the 'profit' of the business. When he recruits someone to work in the business, for a wage rather than a share of the profits, the shopkeeper takes the risk that the extra profit generated by the assistant will at least cover the wage. At that point he starts to think of profit as being the surplus left after he has paid not only his suppliers but his employees.

In a big business the word profit is rarely used to include wages and other employment costs. True, large retail businesses look at their 'gross profit' which includes wages, etc. But their shareholders are more interested in the net profit. In large manufacturing firms the term gross profit is rarely used nowadays. Even when it is, it usually means the difference between sales income and not only purchases but also direct wages. Profit in a large manufacturing business usually means net profit after paying all the suppliers and employees. It may or may not include depreciation and interest charges but it certainly excludes all wages and other employment costs.

Thus in the very small family business, profit represents the income of the proprietors who are also the workers. In the big business profit represents the income of the legal proprietors, most of whom do not work in the business. Profit has become a divisive objective. Those who work in the business may feel aggrieved. Rightly or wrongly, they believe that profit could be put to better use to increase their wages. But the shareholders may feel, rightly or wrongly, that wages are absorbing too much of the gross profit, leaving insufficient net profit to cover the risk.

The investors' view is that business is about profit. The employees' view is that business is about wages, working conditions and job security. At first sight these views seem to be in conflict. And the history of the last 200 years seems to portray a battle between capital and labour. At first, so the story goes, the capitalists had the upper hand. They treated labour as a factor of production. They exploited people who worked hard for long hours and low wages. Then the workers started to fight back. They combined in trade unions to battle for better wages and working conditions. Now, the story seems to say, the workers through their trade unions have the upper hand. The capitalists are receiving their just desserts in the form of lower profits. The state is taking over the ownership of the assets.

But is there really a conflict of interest between capital and labour? Is there any way of reconciling the apparent conflict of interest?

Before trying to answer these questions we have to consider another important group of people, namely, the customers. Indeed their viewpoint is paramount. Customers can do without any one business. But no business can exist for long without customers. A business can only succeed if it meets the needs of most of its customers. Of course, some businesses are more competitive than others. Customers can shop around for food and clothes. They can't shop around for postal and telephone services, electricity, gas and water. But no business has completely compulsory customers. Sooner or later, the customer has the final choice.

It is perhaps a sad reflection of business performance that the consumer has become more vociferous. Many western nations now have consumer associations and consumer spokesmen. Pressure from privately organised consumer groups has led governments to set up consumer services. Legislation to protect the interests of consumers has been introduced. Businesses are no longer allowed to exploit the hapless consumer to the extent that they have done in the past. The employees joined together in trade unions to fight the capitalists. Now the consumers have joined together to fight for their rights. We may yet see the day when the consumer bodies are as powerful as the trades unions and employers' associations.

So the conflict is not just between labour and capital but between the customer and business. Customers want lower prices and better value. Capitalists want higher prices and bigger profits. Workers want higher wages and shorter hours. The conflict is portrayed in Figure 4.1. The opposing viewpoints are represented as the points of a triangle.

And that is not the end of the story. There are other groups of people with an interest in what goes on in business. There are the suppliers who are looking for stable markets and long-term contracts. There are the environmentalists and conservationists who are concerned about pollution and other detrimental effects of business. Above all, there are governments, both central and local, that see business as a source of taxes and rates to finance the non-marketed activities of the state.

Is the conflict of interests inevitable? Which group will win in the end? Human nature thrives on drama. People love stories of battles, contests and games with heroes and villains, winners and losers. The uncertainty of the outcome attracts attention and maintains interest.

But nowadays the distinction between winners and losers in business is blurred. All of us are customers. Most of us are employees. There are very few individual capitalists. Most customers and employees are investors in industry and commerce through pension funds and insurance policies. All tax-payers are investors whether they like it or not.

The simple truth is that in business there is more to be gained from cooperation than from conflict. What business is *really* about is the creation of wealth for the benefit of customers, employees, the providers of capital and society in general. Indeed, this statement applies not just to business in the narrow sense of industry and commerce. It applies to all organisations, whether industrial and commercial companies, public corporations, government services, educational establishments, hospitals, charitable institutions, or social clubs. All of them exist to meet the needs of their customers and employees, the providers of capital and society in general. All of them are in the business of wealth creation.

Some people might object to the idea that wealth creation is the primary objective of all these organisations. Much depends on whether the word wealth is interpreted in a strictly financial sense

Figure 4.1   Conflicting viewpoints of business

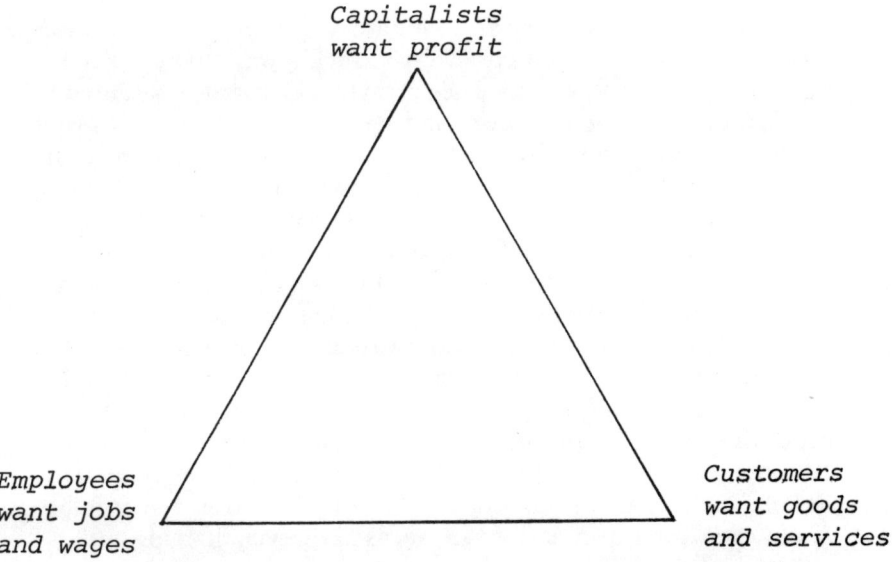

Capitalists
want profit

Employees
want jobs
and wages

Customers
want goods
and services

or whether it is also interpreted in a wider sense to include welfare, well-being and the commonweal. All these words have the same origin. At one time, the words health and wealth were virtually synonymous. Prosperity is not just a matter of material possessions.

If the purpose of business is to create wealth, how should the wealth created be measured? It cannot be measured in terms of sales turnover because that includes the wealth generated by the suppliers. It cannot be measured in terms of profit alone because that excludes the wealth paid out to the employees. The best available measure of wealth created is added value. It is far from perfect as a measure, but it is the best we have yet devised.

One of the limitations of added value is that we have not yet found a way of applying it to non-monetary transactions, e.g. voluntary and charitable work. But the same limitations apply to sales and profit. One of the advantages of added value is that it can be applied to non-profit-seeking bodies.

There are now many organisations undertaking quasi-business activities but with non-profit-seeking constitutions. (There are also some non-profit-making companies that set out to make a profit but don't succeed.) But non-profit-seeking bodies should be just as concerned as commercial enterprise about the efficient use of resources. Some of them show qualities of initiative and enterprise, attention to the needs of their customers, concern for their employees and stewardship of the assets — qualities which are usually thought to exist only in business enterprises. Of course, some non-profit-seeking bodies do not show these qualities. But that is true of some private and public companies.

However, there is an essential difference between the private sector, including quoted companies, and the public sector. Businesses are expected to arrange their affairs in such a way that the needs of their customers are met at a price greater than, or at least equal to, the cost of providing the goods and services. They can incur losses for a few years if they have adequate reserves. But in the long run they must make a profit or at least break even.

By contrast, a publicly owned organisation can be operated at a cost exceeding its income if society feels that this is desirable. Examples are found in nationalised railways, coal mines, steel mills, shipyards, and in various government bodies and local authority services. What matters is not who owns and controls the assets but how effectively the resources are managed by those in charge. There is a popular misconception that nationalised bodies cannot be efficient because there is no yardstick for measuring performance. If they incur losses, the managers are assumed to be incompetent, despite the fact that similar bodies in other countries run at a loss. If they generate profits, they are accused of over-charging and exploiting the consumer. Yet in recent years we have seen examples of large quoted companies being subsidised out of public funds to keep them in business. Economic and social objectives override the purely commercial considerations.

Thus the difference between privately owned companies and public bodies is not as fundamental as it appears at first sight. And the gap is narrowing year by year. What, then, do all these enterprises have in common? Surely it is the efficient use of resources in meeting the needs of their customers, their employees and the providers of capital. The criterion of profit is too narrow as a measure of performance. The only final test is survival. If an organisation fails to meet the needs of its customers, employees and providers of capital, it will not survive, (though some take an unconscionable time to die). No organisation can afford to be static and complacent. It must develop new products, new services, better methods, lower costs, higher wages and better value for the customer.

Thus it can be recognised that the creation and improvement of added value is a primary objective of every organisation. Sales turnover has no merit in itself. Profit alone is not enough. What matters is the amount of added value created by the efforts and ingenuity of the employees in using the assets at their disposal. Added value is the best available measure of the worth placed by the customer on the goods and services supplied by an organisation.

All productive economic activity should aim to add value to materials by using the skill and efforts of people coupled with capital resources in the form of machinery and buildings. By concentrating on this basic objective, an organisation generates income for its employees and for the providers of capital. By adding value a business creates wages, salaries and dividends. Adding value is therefore the primary purpose of every business. A school, a hospital or any publicly owned service should aim to generate wealth, though it may not be easy to calculate the added value created by such organisations.

Unlike profit, which has emotional connotations, added value can be seen by employees and trade unions as a worthwhile objective. The enlargement of added value can be recognised as an essential prerequisite of any increase in wages and salaries. Wealth must be created before it can be shared out. The creation of added value is a valid aim for all organisations. The greater the value of the goods and services produced by the manpower and capital employed in an organisation — industrial, commercial or non-commercial — the better off will be the community as a whole.

4.2 Understanding industry and commerce

Unfortunately, the concept of wealth creation is not understood by many of the people who work in industry and commerce. This is not surprising in view of the fact that, until recently, most companies have not mentioned added value in their published accounts. The chairman's report tends to lay far more stress on the profit generated than on the wealth created. But then, the chairman is reporting to the shareholders whose main interest is profit. It is only in recent years that some companies have started to talk about their wider role in society.

The misunderstandings of the nature and purpose of business and industry are rife. It is difficult for many of those who work in business enterprises to appreciate the significance of their activities. But it is even more difficult for many people outside the business community to comprehend the function of industry and commerce. This is especially true in the world of education.

We cannot entirely blame schoolteachers and college lecturers if their pupils and students emerge from the educational establishments with little or no understanding of business. After all, few teachers and lecturers have themselves worked in industry and commerce. And even if they have, they may have left because they were disenchanted with what they found there. Teachers themselves need to be taught to understand the nature and purpose of business before they can teach it to their pupils.

Fortunately, the gap — indeed gulf — between industry and education has been recognised at government level. Steps are being taken to bridge the gap but this will take years rather than months. A massive programme of education is needed to deal with the problem.

Equally, there are misunderstandings in industry about the nature and purpose of eduction. This is also true about other non-marketed activities financed out of taxation. The medical service and various health services are seen as a financial drain on industry and commerce. There are calls for cuts in education, health and other services in order to reduce the burden of taxation.

These attitudes show a confusion of thought about the business of wealth creation. With the exception of tax collectors and a few other categories, the people who work in the public sector *should* be endeavouring to improve the standard of living of their fellow citizens. The fact that their activities are financed through taxation rather than the market mechanism does not mean that they are not wealth creators. If people paid directly for education and health services and received 'free' food, clothing and other goods, the illusion would be reversed.

The problem with the services financed through taxation is that the customer has too little influence over the quality and cost of the service. With the market mechanism, the customer can shop around, choosing the best buy. The added value created by industry and the marketed services reflects the worth that the customer places on goods and services provided. But in the non-marketed sector, it is not easy to measure the added value because customers cannot compare the value received with the taxes paid. Often the customers have no choice over which school their children should attend or which hospital they must go to.

This is not a plea to change the whole basis of financing education and health services away from taxation to the market mechanism. There are advantages and disadvantages in both methods. It is rather a plea for clearer thinking about the business of wealth creation. Industry and commerce are not the only sectors of the economy that create wealth. There are wealth-creating services in the public sector too. The big problem is that of measuring their contribution and giving the customer better value for money.

However, the first step must be to bring about a better under-standing within industry and commerce of the purpose of business in terms of wealth creation. This means that the right kind of information must be communicated to employees, shareholders, customers, government, suppliers and other interested parties. Conventional methods of presenting accounting information have hindered rather than helped such communication. New methods must be found to meet the wider objectives. Added value offers a sound basis for better communication.

5 ADDED VALUE

and accounting information

5.1 The profit and loss account

Every limited company is required by law to submit to its share-holders an annual report and accounts. It must provide certain information stipulated in the various Companies Acts. Many progressive firms give far more information than the legal minimum. As a precaution to ensure that the accounts present a true and fair view of the financial position of the firm, the books and accounts must be audited annually by a recognised firm of auditors who must employ qualified accountants.

The historical function of the accountant in auditing is that of stewardship. He acts as a watchdog for the shareholders. He there-fore tends to take a cautious and conservative view. The main purpose of the first Companies Act was to protect the interest of the shareholders. The auditor viewed the directors with some degree of suspicion. Conversely the directors came to regard auditors as a necessary evil, or at least an obligatory expense, forced upon them by the requirements of the Companies Acts.

Inevitably a barrier grew up between directors and auditors. The former resented the intrusion of the latter who tended to become secretive. Accountants developed their own language and special terms, just as lawyers, doctors and other professions have done. Eventually the attitudes of both sides could be summed up in the words of R.G.A. Boland, who dedicated some of his books to:

Accountants who, in the game of business, aspired to be players or at least umpires, but were relegated to the humble office of scorekeepers. Their revenge for this ignominy was to keep the score in such a way that neither the players nor the umpire could ascertain the true state of the game.

It is a pity, if not a tragedy, that this cold war between auditors and directors lasted so long. Fortunately, the barriers are fast being broken down. But there is a long way to go. The vast majority of shareholders do not really understand the financial information presented in the glossy annual accounts. Many company directors, other than those with accounting qualifications or some training in financial matters, do not fully comprehend the audited accounts. Certainly, far too many middle managers and supervisors are quite incapable of interpreting the information in the annual accounts. No wonder, then, that employees, especially those on the shop floor, are hazy about the real meaning of the figures. No wonder they sometimes misinterpret the information.

The traditional method of presenting the profit and loss account does not help the uninitiated to understand its mysteries. Consider the example illustrated in Figure 5.1. This shows the group profit and loss account for ICI, as published in the annual report and accounts.

The first point to note is the enormous information gap between the sales and profit figures. The table does not show what happened to the difference between the sales income and the trading profit figures. In 1975 the company received £3,129 million from its customers. Approximately 90 per cent of that money disappeared somewhere, leaving the other 10 per cent as trading profit.

The Companies Acts of 1948 and 1967 did not require companies to publish full details of their costs. In particular, there was no requirement to publish the figure for purchases of raw materials and bought-in services. Perhaps it was felt that such information was not essential for shareholders. After all, they were supposed to be interested only in profit and dividends. Perhaps it was felt that such information might be damaging to the company, especially if competitors had access to it. The latter point might be valid for some small firms. But for a giant company with diverse interests, it seems unlikely that such knowledge would give competitors a big advantage. In any case, there is a time lag between the end of the financial year and publication of the results. And if competitors really want to know, there are other means of finding out.

However, ICI was one of the progressive companies that did publish the figure for purchases of materials and services. If a shareholder or employee or any other interested party really wanted

Figure 5.1 ICI: group profit and loss account

For the year ended 31 December	1975 £m	1976 £m
Sales to external customers	3,129	4,135
Trading profit	319	519
Exchange gain (overseas subsidiaries)	29	58
Investment income	39	53
Interest payable	(73)	(93)
Employees' profit-sharing bonus	(17)	(27)
Share of profits of associated companies	24	30
Profit before taxation and grants	321	540
Taxation less grants	(125)	(214)
Profit after taxation and grants	196	326
Extraordinary items	(1)	(47)
Profit after taxation and extraordinary items	195	279
Applicable to minorities	(24)	(34)
Applicable to parent company	171	245
Dividends	(59)	(83)
Profit retained for year	112	162

to know what happened to most of the sales income, he or she could find out by looking in the accounts for the information. From 1956 to 1971, the figures were shown at the front of the annual report in a table headed 'Disposition of Group Income'. Then in 1972 and 1973, the purchases figure was incorporated in the text of the report. In 1974 the Disposition of Income table was resumed. Then in 1975 and 1976 the report included a statement of added value.

ICI was exceptional in giving such detail of its costs. A study of the published accounts of 100 large British companies showed that, before 1975, most of them did not publish a figure for purchases. In some cases the figure could be deduced by subtracting from the sales turnover the sum total of wages and salaries, depreciation, interest charges and profit before tax. But in many cases this calculation did not give the answer. Either the company did not show the remuneration of overseas employees or there were accounting complications caused by such items as government grants, investment income, profits from associated companies and other miscellaneous items. Thus ICI was one of the few enlightened firms that reconciled the figures. Even so, its conventional profit and loss account left a big gap between sales and profit.

The next point to note is the variety of profit figures given in the table. There is trading profit, profit before taxation and grants, profit after taxation and grants, profit after taxation and extraordinary items and, finally, profit retained. Which of these figures should the employee or shareholder take as the guide? The employee might look at the trading profit of £319 million in 1975, and note that it increased by 63 per cent to £519 million in 1976. He might then divide these figures by the approximate number of employees in the group, i.e. 195,000 in 1975 and 192,000 in 1976. If so he would find that the trading profit per employee rose from £1636 to £2703. He could perhaps be forgiven for jumping to the conclusion that his own remuneration could be increased. He might accept that the tax collector has a prior claim on the profit. But in 1976 there was still nearly £1700 of profit per employee after taxation. He might accept the need for the extraordinary items (mainly provision for restructuring the man-made fibres business). He might agree to the payment of dividends. But there was still the matter of £162 million retained profit in 1976. That represented £844 per employee.

Similar figures and similar interpretations could be put forward for many companies, large and small. Employees and customers might look at the published figures and believe that there was scope to increase wages or reduce prices.

The shareholder might take a different view. The dividend from ICI and many other companies represents little more than the investor might have received from safe securities, in some cases less. In theory, the retained profit belongs to the shareholder. In practice, the shareholder can only benefit if the retained profit is used to generate more profit from which a bigger dividend is paid.

Another vital point to note about most large companies is that the profit figures run into millions of pounds, indeed tens or hundreds of millions. The individual employee can be excused for thinking that if his or her remuneration were doubled, it would make very little difference to the total profit. When the profit figure has six noughts on the end of it, most employees have difficulty in putting it into perspective.

There are other aspects of company accounts that are open to misinterpretation by employees and shareholders. For example, in the balance sheet there is usually an item called 'reserves'. Many employees and many private shareholders might be under the misapprehension that these 'reserves' represent cash put away for a rainy day. In reality, the reserves are either retained profit or the appreciation of the value of land and buildings. These millions of pounds are rarely available as cash. They are tied up in the fixed assets of land, buildings, plant and equipment or in stocks and debtors Any significant amount of cash in the company will be earning interest until it is needed to finance more stocks and debtors. But the employees and the small shareholders can easily misunderstand the meaning of the term 'reserves'.

Thus it is not surprising that the conventional method of presenting accounting information in the annual report is open to misinterpretation. The institutional investors, such as insurance companies, have their own experts who know what the figures mean. The private shareholders and most employees are not experts in accountancy. They need some guidance to avoid misunderstanding.

Much has been done in recent years to take some of the mystery out of accounting reports. There are several books that explain accounting terms in layman's language. Many companies issue special employee reports that set out the facts, not just in tables of figures, but in the form of charts and diagram. Some companies have prepared films and audio-visual presentations to help to explain the accounts. The Confederation of British Industries issued a booklet in 1974 outlining methods of presenting company accounts to non-experts.

5.2   The added value statement

Probably the biggest single advance in recent years in the presentation of accounting information is the introduction of the value added statement. A few companies have published added value figures for several years. But the impetus for more general disclosure came in a discussion paper prepared by the Accounting Standards Committee. Their paper, entitled *The Corporate Report*, was published in July 1975. It recommended various changes in corporate reporting including such ideas as an employment report showing details of numbers of employees, age distribution, sex, geographical location, hours worked, remuneration, time spent on training, names of trade unions, labour turnover, absenteeism, accidents, etc. On financial reports it recognised the need for additional statements such as money exchanges with government, transactions in foreign currency, disaggregation between separate divisions in large companies and a statement of future prospects. In particular, it recommended the publication of a statement of added value showing how the benefits of the efforts of an enterprise are shared between employees, providers of capital, the state and reinvestment. The Committee went on to say:

> The simplest and most immediate way of putting profit into proper perspective vis-a-vis the whole enterprise as a collective effort by capital, management and employees is by presentation of a statement of added value (that is, sales income less materials and services purchased). Added value is the wealth the reporting entity has been able to create by its own and its employees' efforts. This statement should show how value added has been used to pay those contributing to its creation. It usefully elaborates on the profit and loss account and in time may come to be regarded as a preferable way of describing performance.
>
> There is evidence that the meaning and significance of profits are widely misunderstood. It is not the purpose of this report to attempt to justify the profit concept. We accept the proposition that profits are an essential part of any market economy, and that in consequence their positive and creative function should be clearly recognised and presented. But profit is a part only of value added. From added value must come wages, dividends and interest, taxes and funds for new investment. The interdependence of each is made more apparent by a statement of added value.
>
> We recommend that business enterprises, and where appropriate other economic entities, include as a part of their corporate reports and in a prominent position, a statement of added value containing as a minimum the following information:
>
> *a*   Turnover.
> *b*   Bought-in materials and services.
> *c*   Employees' wages and benefits.
> *d*   Dividends and interest payable.
> *e*   Tax payable.
> *f*   Amount retained for reinvestment.
>
> The statement of added value provides a useful measure to help in gauging performance and activity. The figure of value added can be a pointer to the net output of the firm; by relating other key figures (for example, capital employed and employee costs) significant indicators of performance may be obtained.

Following the publication of *The Corporate Report* some companies included an added value statement in their annual reports for 1975. In May 1976 the Department of Trade issued a preliminary draft paper *Aims and Scope of Company Reports* recommending, amongst other things, that company reports should include a statement of added value. Then in July 1977 a Green Paper was published on *The Future of Company Reports*. This supported many of the ideas put forward in the previous documents, including the added value statement. The Green Paper (Cmnd 6888) said:

There is a growing tendency for listed companies to prepare statements of added value, either as part of their audited accounts or more frequently as a supplement to them. In the Government's view, the added value statement is a useful addition to the financial information produced by companies. A company's turnover is in part the result of other people's work, namely the raw materials, products and services which the company has purchased from outside, and in part the result of the efforts of the company's workforce and the use of its physical and financial assets. This latter part is the value added by the company, and can therefore be expressed as turnover less goods and services purchased from outside.

Having identified the added value, it is then of interest to show how it has been distributed by way of payment to employees, to those who have provided the capital by way of dividends on share capital and possibly interest on loan capital and to Government through taxation; and indicating the balance retained in the business.

Care is needed in using added value as a measure of wealth creation, and in interpreting changes in added value. But with this proviso, the identification of added value provides a useful indication of performance, to supplement other financial information.

The information required to produce an added value statement is readily available to companies. Preparation of the statement calls for the re-ordering of information already revealed in the profit and loss account, together with disclosure of certain information which does not at present have to be published. The Government has concluded that there should be a legislative requirement that companies should publish such statements. The Government envisages a requirement framed in general terms, with the detailed rules on the method of preparation and form being suitable matters for prescription by Accounting Standard; this would provide flexibility for companies such as banks or insurance companies where at present there are special disclosure requirements, and where there may be a case for different forms of added value statements. However, in order to provide a measure of consistency in preparation, the Government believes that added value statements should be subject to audit.

It now seems certain that legislation will be introduced requiring companies to publish added value statements. Ten years previously similar suggestions were viewed with considerable suspicion by businessmen. Now the idea is seen as helpful rather than harmful.

So what does a statement of added value look like? An example is given in Figure 5.2, as shown in ICI's annual report, for the same years as the profit and loss account in Figure 5.1. It follows a similar format to that recommended by the Accounting Standards Committee.

The main virtue of the statement of added value is that it shows what has happened to all of the income from sales. There is not the huge gap that exists in the profit and loss account. We can see at a glance that well over half of the sales income went to the people supplying materials and services to the company. In fact the proportion was nearly 60 per cent. The statement of added value also identifies the wealth created by the company whereas the sales turnover includes the wealth created by its suppliers.

The section showing the disposal of total value added reveals that well over half of the added value went to the employees. Again the proportion is nearer to 60 per cent. Thus it is quite clear that the employees were receiving the lion's share of the added value.

Next, the statement of added value shows the proportion allocated to the government. However, this includes deferred tax which is actually retained in the business and is unlikely to be paid out. But the ordinary shareholders of ICI received only about 8 per cent of the added value. The remainder of the added value was re-invested in the company either as depreciation or as retained profit or as deferred tax.

This method of presenting accounting information is much better than the conventional profit and loss account. It tells the whole story instead of only a small (but vital) part. It leaves no huge unexplained gaps. The facts are laid bare for all to see. Moreover, it puts profit into perspective as only a small part of the wealth created. Indeed, there is no figure in the statement of value added that corresponds with the profit figures shown in Figure 5.1. Instead, the profit is broken down into its various components of employees' bonus, interest charges, taxation, dividends and retained profit.

Although many companies have issued added value statements in a format similar to ICI there are variations on the theme. One example is given in Figure 5.3 showing the added value statement for Record Ridgway Limited. This starts in a conventional way, showing sales, purchases and added value. But the section showing the disposal of the added value is different. Instead of grouping the items into four categories it simply lists them in sequence. More important, the

Figure 5.2 ICI: sources and disposal of added value

		1975 £m		1976 £m
Sources of income				
Sales		3,129		4,135
Royalties and other trading income		30		40
Less: materials and services used		(1,842)		(2,458)
Value added by manufacturing and trading		1,317		1,717
Share of profits of principal associated companies and investment income		63		83
Exchange gain on net current assets of overseas subsidiaries		29		58
Extraordinary items		(1)		(47)
TOTAL ADDED VALUE		1,408		1,811
Disposal of total added value				
Employees			833	1,020
Pay, plus pension and NI contributions	816		933	
Profit-sharing bonus	17		27	
Governments - corporate taxes less grants			125	214
Providers of capital			156	210
Interest paid on borrowings	73		93	
Dividends to stockholders	59		83	
Minority shareholders in subsidiaries	24		34	
Reinvestment in business			294	367
Depreciation set aside	182		205	
Profit retained	112		162	
			1,408	1,811

The form of presentation differs slightly from that given in the 1975 Annual Report. One change is the identification of the added value arising from manufacturing and trading, rather than other sources of income. Another change is that the employees' portion of the added value includes the employers' national insurance contribution. There are also some changes in the treatment of government grants and exchange gains. These changes make it difficult to compare figures with earlier years penultimate item is entitled, 'To finance inflation'. This represents the sum of money that would be needed if the stocks of raw material, work-in-progress and finished goods were valued at current cost instead of historical cost. The balance of the retained profit is then shown after the allowance for inflation. This format helps to put profit into proper perspective, especially after adjustment for inflation.

Some companies have adopted slighty different ways of presenting the information. Instead of showing just corporation tax as the amount going to the government they have shown also the income

Figure 5.3 Record Ridgway: added value statement

	1976 £'000	1975 £'000
Turnover	15,523	11,993
Less materials and services	8,042	6,159
ADDED VALUE	7,481	5,834
Applied as follows:		
To employees - Wages salaries and pensions	5,141	4,181
To government - Tax on profit	721	456
To bankers - Interest on overdrafts	71	149
To Record Ridgway shareholders - Dividends	271	234
To Australian shareholders - Profit entitlement	42	23
To replace assets - Depreciation	219	139
To finance inflation - Maintenance of working capital	875	614
To finance expansion - Profit retained	141	38
ADDED VALUE	7,481	5,834

tax paid by employees. It is argued that this method reflects more accurately the government's share of added value. However, if this corporation tax includes deferred tax it is misleading to say that it has all gone to the government. So some firms have shown the deferred tax as retained in the business. Other companies have included the national insurance contributions of both employees and employer in the portion going to the government. One retail business even included the whole of the VAT they collected over the counter in their total sales and thus showed an even bigger share going to the government. Finally a large brewery business included the customs and excise duty in the share to the government.

There is some merit in these formats, especially in terms of showing how much of the added value goes to the government. But if such formats were to become general practice, it would be better to show the items separately. Income tax can vary from one firm to another and from year to year. If it were not separated from corporation tax it would be impossible to assess the reason for changes. More important, if VAT is to be included it should be only the net amount handed over after allowance for VAT paid to suppliers. And the VAT and other customs and excise duties should be shown separately from income tax and corporation tax.

5.3 Presenting accounting information

Of course, simply writing the figures down on a piece of paper is only the first step in communicating accounting information. Much more must be done. The information must be presented in ways that facilitate understanding. Many people find that tables of figures make for dull reading. It is easier to assimilate the information if it is presented in the form of charts and diagrams. Even the simple process of converting the data in Figure 5.2 into the block diagram given in Figure 5.4 helps to show the relative magnitude of the different items.

Many companies have developed special versions of the annual report for employees. Some of them take the form of a newspaper with photographs of the products, the plant and the people. Most of them make extensive use of diagrams and charts for presenting financial information. They use block diagrams, graphs, pie charts, flow charts, piles of coins and other devices to illustrate the figures. They use different colours to highlight the main points. They show where the money has come from and where it has gone. They show the company's assets and how they are financed. No expense is spared in endeavouring to present the information in attractive, easy-to-read ways.

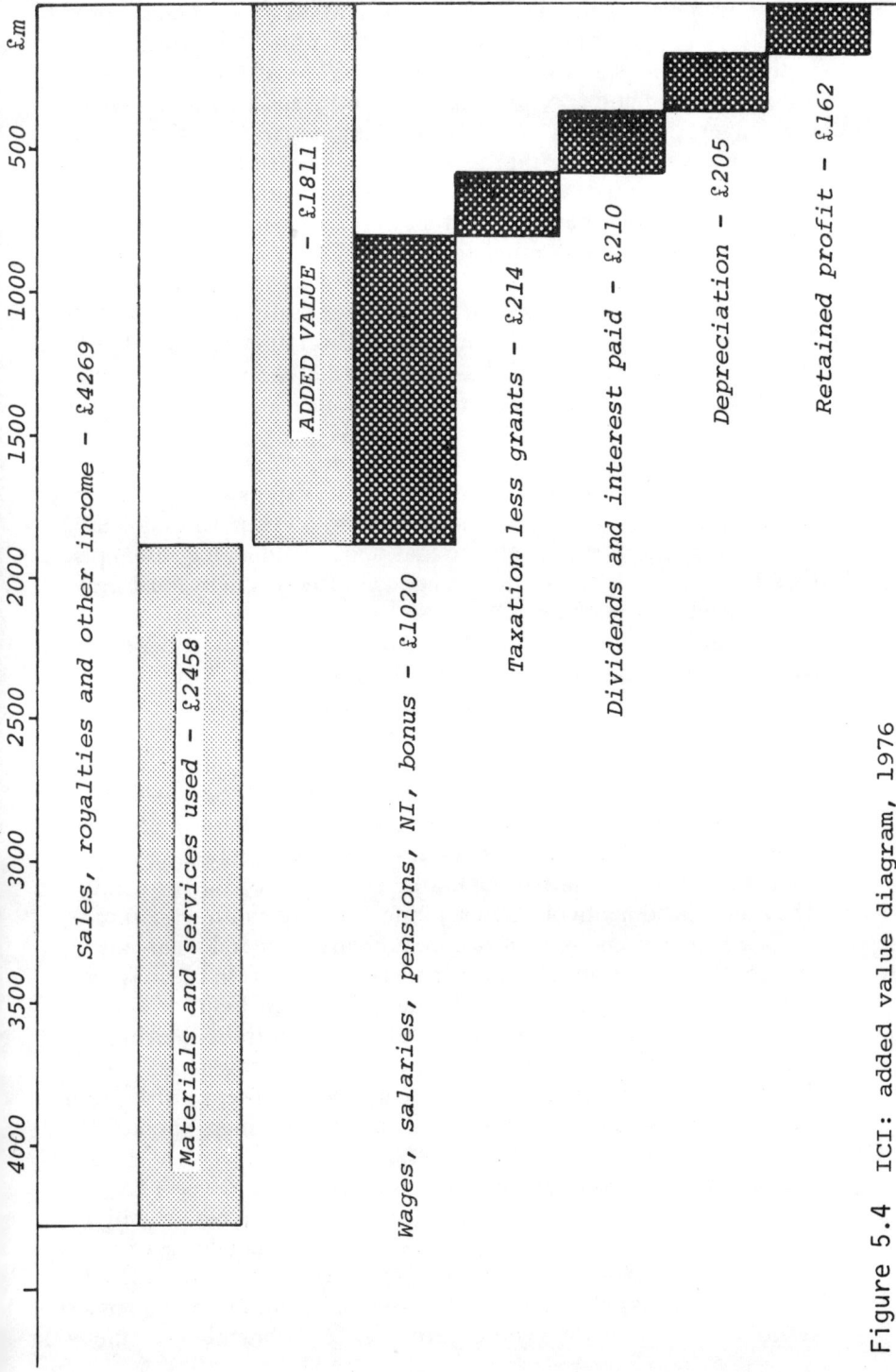

Figure 5.4 ICI: added value diagram, 1976

But even converting figures to charts and diagrams is only one step in the process of communication. Presenting the information in these ways does not guarantee that it will be assimilated, much less understood and accepted. Glossy reports and multicoloured newspapers can finish up in the waste bin, unread or misinterpreted. The right way to ensure that this does not happen is to supplement the written information with verbal communication.

Some companies have felt that all that is necessary is to bring groups of employees together for a few minutes so that a manager can explain the figures. He then invites questions and comments. On finding few if any questions, he assumes that the job is done or that employees are not really interested. But understanding does not come from reading and listening. If it did, then education could be left to the newspapers and television. There would be no shortage of experts amongst readers and viewers.

True understanding starts when the communication process is two-way. The information must not only be expressed in terms that can be understood by the reader or listener. Steps must be taken to stimulate a response to ensure that understanding has taken place. Questions and comments are not enough. The recipient must be encouraged to participate.

You can't learn to swim without getting into the water. You can't learn to drive a car without sitting behind the wheel. You can't learn to write without putting pencil to paper. You can't learn to interpret accounting information without getting involved.

This means that if companies really want their employees to understand what the business is about and what the accounting information means, they must take more trouble than simply publishing employee bulletins and holding brief meetings. Some companies have recognised the nature and depth of the problem. They have become involved in genuine two-way communication.

There are various ways of tackling the problem and what suits one company may not suit another. But a description of the way that one large company set about the task may be helpful. They set up a project team of line managers, accountants, training officers and other specialists to devise a communication programme for their 50,000 employees. Ultimately, the team created a series of 'learning packages', each covering a particular aspect of business operations and financial information.

Each 'module' consisted of a 15-min video cassette recording which could be presented to small groups of employees, usually 10-15 people. Each package also included an associated task or exercise to be carried out by small groups within the group. In addition time was allowed for discussion and a question and answer session, with the 'tutor'. These 'tutors' were not members of the training staff but line managers and supervisors. The whole 'course' occupied about 2½ days but could be broken down into shorter periods.

Clearly this programme of communication involved a massive investment by the company. The cost of preparing TV programmes to professional standard was not cheap. But that cost was small compared with the cost in wages, salaries and expenses for the people involved. A plan of such magnitude was not undertaken lightly. It had commitment at board level and wide managerial backing.

Was it worth it? This question is difficult to answer because the long-term results may be different from the short-term effects. To some extent, the project was an act of faith. Nevertheless, some benefits were clear. In one works, a difficult reorganisation plan went ahead smoothly after the communications exercise. The project stimulated many questions and suggestions from employees, demonstrating a desire to know more and to participate.

However, the response was not entirely favourable. Some participants criticised the practical exercises which involved mathematical tasks. But it would be impossible to convey an understanding of business without at least some simple arithmetic.

Other companies have pursued similar programmes, using material developed by their own staff or by outside specialists. The secrets of success seem to lie in careful preparation of the material presented, deliberate steps to involve the participants in applying the information, and ample opportunity for discussion with a leader who can give straight answers.

Of course, there can be no guarantee that such measures will work easily in every business. Where there is a long history of poor industrial relations, such steps may be seen as yet another attempt by management to hoodwink the employees. Revealing accounting information to those who are deliberately seeking opportunities to disrupt the company's activities may be simply offering ammunition to the enemy. There is no easy way to overcome deep-seated and long-rooted problems of mistrust. But a start must be made somewhere. The road may be hard and long. Yet the alternative may be far worse. The vicious circle of suspicion and opposition must be converted into a virtuous circle of enlightenment and cooperation.

6 ADDED VALUE

and wages and salaries

6.1 The link between added value and pay

Added value is a measure of the wealth created by a business, an industry or a country. Wages and salaries are part of the added value, usually the major part. The remainder of the added value consists of depreciation, certain expenses (e.g. rent and rates), interest charges and profit. This simple statement is illustrated in Figure 6.1.

From this diagram it is clear that if the share of added value going to the employees increases, there may be insufficient left over to finance the cost of capital. If this happens in a particular company, the business will find it difficult to attract and retain funds for investment. Sooner or later, it will go out of business, unless subsidised.

Conversely, if the share of added value going to the employees decreases, the business will show an increase in profit. It will be able to attract and retain more funds for investment. If the profits are high, other companies will be attracted into the market until prices and profits fall. But also when the profits rise, the employees will seek increases in wages and salaries. Sooner or later they will lift their share of the added value to the point where the reward to capital is just sufficient to keep the company in business.

Thus there is a fundamental link between added value and wages and salaries. Businesses that become unprofitable, because their wage and salary bill absorbs too high a proportion of the added value, either go to the wall or go to the government. Businesses that offer new products and services, which generate good profits after the payment of competitive wages and salaries, survive and prosper.

In a free market economy, some firms decline and others expand. Some firms go out of business and new firms start up. The mechanism that determines survival and growth is the relationship between added value and wages and salaries. In any one business the ratio may vary from time to time. But over the whole economy, or over a large number of businesses in different industries, there is a tendency towards equilibrium.

6.2   Added value and pay in the USA

This phenomenon is not new. It has been studied and reported upon for at least 50 years. An article appeared in the *American Economic Review* in 1928 by the University of Chicago economist, Paul Douglas, and his colleague, Charlie Cobb. They attempted to explain the relative constancy of the shares of labour and capital in the national income. Douglas spent the better part of 20 years developing the Cobb-Douglas production function which measured the relationship between output, labour and capital.

In 1932-33, another American economist, Alan Rucker, showed that the ratio of wage costs to added value in the USA manufacturing industry had been virtually constant from 1899 to 1929. His work formed the basis for the so-called Rucker Plan, a group bonus scheme based on added value.

Figure 6.1   Added value and wages

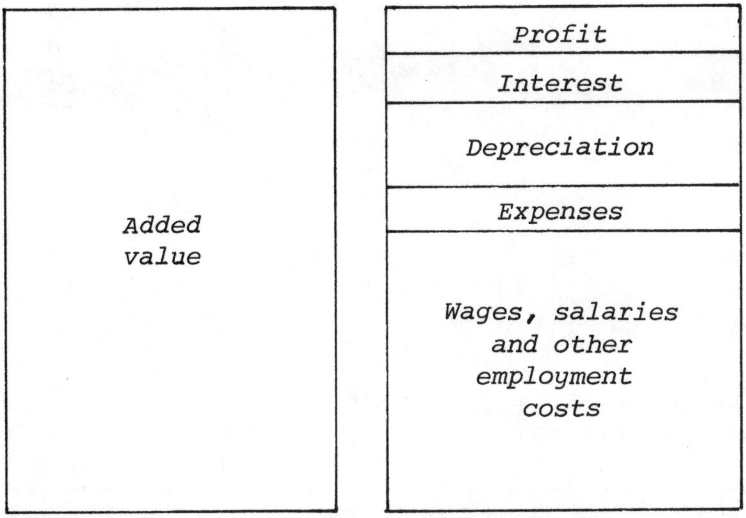

6.3 Added value and pay in the United Nations

In 1969, P.J. Loftus, the Director of the Statistical Office of the United Nations, published a paper showing wages and salaries as a percentage of added value in the manufacturing industry in various countries. A summary of the results is shown in Figure 6.2. Within most countries the ratio is fairly stable. The differences between the countries can be explained partly in terms of differences in the definition of added value and wages and salaries. However, there is little doubt that Japan, with the lowest ratio, had a higher percentage of added value available for investment. This helps explain the high rate of productivity increase in Japan in recent years.

Loftus also quoted figures for 20 different industries in 14 countries. With few exceptions the wage/salary ratio was lowest in the tobacco industry and highest in the vehicle industry. In most countries, the chemical industry showed a low ratio, as might be expected in a capital-intensive industry. Conversely, in most countries, the clothing industries showed a high ratio, as might be expected in a labour-intensive industry.

Loftus also gave figures for the USA manufacturing industry from 1889 to 1965. These are shown in Figure 6.3. The stability of

Figure 6.2 Wages and salaries as percentages of added value in manufacturing industry*

Country	1953 or (1954)	1958	1963 or (1964)
Australia	58	54	52
Canada	50	50	49
Finland	(54)	52	51
Ireland	53	53	51
Japan	40	41	37
Netherlands	46	51	46
New Zealand	59	59	54
Norway	45	50	51
Rhodesia	50	53	50
South Africa	47	48	45
Sweden	58	57	57
United Kingdom	(55)	57	53
United States	(54)	52	49
USSR	55	45	(45)

*Source: P.J.Loftus, *Lloyds Bank Review* (April 1969)

Figure 6.3 Wages and salaries as percentages of added value in US manufacturing industry

Year	Ratio	Year	Ratio	Year	Ratio
1889	54	1933	45	1956	53
1899	47	1935	52	1957	54
1904	50	1937	51	1958	55
1909	50	1939	52	1959	53
1914	54	1947	53	1960	54
1919	52	1950	53	1961	54
1923	53	1951	55	1962	53
1925	51	1953	56	1963	52
1927	50	1954	56	1964	52
1929	47	1955	53	1965	51

the ratio is remarkable. Through boom and slump, war and peace, the ratio averages 52 per cent, with a range from 45 to 56 per cent.

6.4 Added value and pay in Britain

In 1939, Lord Keynes, the economist, commented on the stability of the proportion of the national income accruing to labour in both Britain and the USA. In June 1952 the *Economic Journal* carried an article by Professor E.H. Phelps-Brown and P.E. Hart quoting figures for the distribution of home-produced national income of the UK. They showed that, from 1870 to 1950, wages as a percentage of the national income had been virtually constant at about 40 per cent. Their analysis also showed that salaries as a percentage of the national income rose from 16 to 23 per cent whilst the shares of profit and rent declined. Figure 6.4 shows the data for every fifth year. It is important to note that although the ratio of wages to added value was stable, the number of wage earners as a percentage of the employed population fell from 84 per cent in 1870 to 66 per cent in 1950. Thus the standard of living of wage earners rose faster than that of salary earners. Obviously there must be a limit to such a trend.

A detailed study of the British *Census of Production,* carried out by E.G. Wood, showed that the ratio of wages and salaries to net output has been fairly stable for many years. This is true over the whole of manufacturing industry. In some industries the ratio has tended to rise whereas in other industries it has tended to fall. Figure 6.5 shows the data for all manufacturing industries from 1958 to 1976. Before 1970, the Census was compiled every five years. Since 1970, it has been compiled on an annual basis.

Figure 6.4 Distribution of national income – with figures given as percentages of national income

Year	Wages	Salaries	Rent	Residue including profit
1870	38.6	16.2	14.2	31.0
1875	42.4	14.1	13.4	30.0
1880	39.8	14.9	15.3	30.0
1885	39.8	17.5	15.7	27.0
1890	41.5	17.3	12.9	28.3
1895	40.6	19.0	13.0	27.4
1900	40.7	17.9	11.6	29.8
1905	38.3	19.3	12.6	29.8
1910	37.8	19.2	12.0	31.0
1915 1920	Data not available			
1925	41.8	24.3	7.5	26.4
1930	41.0	25.6	9.3	24.1
1935	40.3	26.1	9.6	24.0
1940	38.3	22.0	7.8	31.9
1945	39.7	22.0	5.9	32.8
1950	41.9	23.4	4.6	30.1

Figure 6.5 Wages and salaries as percentages of net output in UK manufacturing industry

Year	Net output, £m	Wages and salaries, £m	Wages & salaries/ net output, %
1958	7,848	4,454	56.7
1963	10,820	5,704	52.7
1968	15,289	7,768	50.8
1970	18,532	9,737	52.4
1971	19,847	10,533	53.0
1972	21,941	11,249	51.3
1973	26,600	12,858	48.3
1974	32,331	15,673	48.5
1975	36,797	19,271	52.4
1976	44,872	21,832	48.7

From 1973 onwards the definition of net output was changed slightly to include transport costs. In most industries transport costs are small in relation to net output. The effect is to reduce the ratio by about 3 percentage points over the whole of the manufacturing industry. Thus the ratios in the last three years of the table should be increased to about 51.3, 51.5, 55.4 and 51.7, respectively. Thus, it appears that the ratio did not change much between 1958 and 1976. The increase in the ratio in 1975 was perhaps a reflection of the surge of wage increases, granted in 1974, which worked their way into the economy in 1975 when prices were still held down by the Price Commission. In 1976, the balance was restored when wages were restrained and rose less rapidly than prices.

Thus, although the ratio has fluctuated somewhat, the changes are quite small. Despite all the ups and downs of the economy, the booms and slumps, inflation, devaluation, the floating pound, price restrictions, wage freezes, incomes policy, grants and subsidies, taxation, legislation and several changes of government, the relationship between wages and salaries and net output has been remarkably stable.

6.5 Added value as a basis for incomes policy

This phenomenon could be used as the basis for a national incomes policy based on productivity. In theory, there is no need for a national incomes policy. The market mechanism can be left to sort out the equilibrium between wages and salaries and added value. But in practice the market mechanism is distorted. It is restricted by the actions of powerful monopolies, whether employers or trade unions. Some means must be found either to break up the monopolies or to control their power. A formula for reaching agreement on wage and salary settlements is needed.

Wage negotiations are too often conducted on the basis of a trial of strength. One side pleads that the demand for increases cannot be met. The other side argues that the present wage is inadequate, has failed to keep pace with increases in the cost of living, or that other groups have had bigger increases. Eventually, one side or the other gives way. Increases in pay are granted. The customer pays the bill in higher prices, or pay increases are restricted. Then employees become disgruntled and unenthusiastic about their work.

But the added value concept could provide a solution to the eternal conflict. The proportion of the net output going to wages and salaries could be monitored in each industry against norms established on Census data. The incomes policy could be determined by setting a limit to the proportion of the net output to be paid out as wages and salaries. This does not mean a limit to the individual's wage or salary. It does mean that the only way of obtaining an increase in the average wage or salary is either to increase the total net output or to share out the same net output amongst fewer people. This need not mean unemployment if the manpower released is absorbed by other sectors of the economy.

Of course, there would be many problems in implementing such a policy. There are some technical problems in using the Census data. Net output includes certain costs which, strictly speaking, are not part of added value. Fortunately, from 1973 onwards the Census has been compiling figures for both net output and added value. Also, the costs of the employers' contribution to national insurance and pensions have been excluded from wage and salary data. However, the figures are known and adjustment could be made.

Another problem is that in the industries which are becoming more capital-intensive there is a need to reduce the ratio of wages and salaries to net output in order to pay for the investment. And in some industries the ratio has already risen so high that investment is being subsidised by the taxpayer. But in many industries the historical ratios are still a reliable guide.

The Census covers only manufacturing, mining and utilities. There is little data for the non-manufacturing sectors such as retailing, insurance, banking, etc. But figures could be compiled from company data in such sectors. A bigger problem is that there is no corresponding data for the non-marketed sectors such as education, health, local and central government. However, acceptance of the principle in the manufacturing industry would point the way to the need for measurement in other sectors.

But the biggest problems are not in the technical calculations. They lie in the fact that this proposal would cut across traditional methods of bargaining. In many industries wages are settled by bargaining through several trade unions. Salaries in the same industries are the subject of separate bargains. It would not be easy for the representatives of the various groups to come together to reach broad agreement on the method of sharing out the total wage and salary bill available from the formula. However, given time and goodwill, solutions could be found.

An even more difficult problem, though not insuperable, is that of explaining the policy to the individuals and groups in each industry and organisation. As yet, few British managers and trade union leaders, let alone ordinary employees, understand the concept of added value and its importance in connection with wages and salaries. A massive programme of education and training is needed at all levels to get the message across. The simple truth is that wealth must be created before it can be shared out. Thus, stark reality is not as attractive as the illusion that real wages and salaries can be increased without any increase in output.

6.6   Added value for company pay policy

There is no need for companies to wait for the government to lay down guidelines for a national incomes policy based on productivity. Many firms could devise their own policy for wage and salary determination based on added value. Instead of reacting negatively to the inevitable demands for increases in wages and salaries, companies could establish a set of positive principles based on added value. A study of the accounting information available in many companies shows that the ratio of the total wages and salary bill to added value has been virtually constant for many years. Where this is the case, the company can use the phenomenon as a basis for determining the total wage and salary bill that the company can afford without jeopardising its investment programme or passing the burden on to the consumer.

Again there are the problems of traditional bargaining groups within the company. But the smaller the number of people involved, the easier it is to reach agreement. In a large company with tens of thousands of employees the process of negotiation may take longer than in smaller firms with only a few hundred employees. Even so, it would be easier to reach agreement within one large company than across a whole industry or the whole nation.

Again the problems lie not so much in the principles as in the practice. The first essential is that management should understand the implications. They must become familiar with the concept of added value, the methods of calculating the figures, the interpretation of ratios. Similarly, the trade unions and employee representatives must learn how to use added value information, A successful agreement must be based on mutual recognition of the facts and figures, rather than on opinions and prejudices and a trial of strength.

In turn, this means that companies must be prepared to invest in education and training of their managers and employees, including trade union representatives. There is no short-cut route to reach the degree of understanding needed. Nevertheless, the benefits can emerge soon after the first steps are taken along the road.

7 ADDED VALUE

and bonus schemes

7.1. Types of incentive scheme

Although bonus schemes have been and still are widely used in industry, less than half of the working population are involved in them. The most widely used systems of payment are those based on time. People are paid either an hourly rate, a weekly wage or a monthly salary. The advantage of time-based systems of payments is their simplicity. It is easy for the employer and employee to calculate and check the total remuneration. Even these systems offer a form of incentive because there can be an inducement to gain promotion to a higher rate of pay. But their main disadvantage is the lack of connection between output and pay. Well designed incentive schemes can offer the employee the opportunity of higher pay. At the same time they can offer the employer lower costs per unit of output.

The simplest form of incentive scheme is the piecework system. The reward is based on a constant price per piece produced, irrespective of the time taken. Pure piecework rarely exists nowadays. Most pieceworkers have a guaranteed minimum time rate of earnings. Piecework is still used in certain industries and it provides a strong incentive. Remuneration is in direct ratio to the output of the employee.

In practice, piecework systems tend to develop 'tight' and 'slack' prices. This leads to anomalies of earnings between employees. Piecework prices need frequent adjustment for inflation. When improvements in methods or materials make the job easier, piece-work prices cannot be easily adjusted downwards. Some so-called piecework systems incorporate so many adjustments, allowances and complications that they have lost their merit of simplicity. More important, in some schemes the main incentive is to fiddle the figures rather than produce more output for more pay.

To overcome the problems associated with piecework, many other types of incentive scheme have been devised. There are the premium bonus schemes based on time saved'. There are the schemes based on work study using standard times established by systematic work measurement. All have their advantages and disadvantages but many have fallen out of favour. The problems are associated more with misuse and abuse rather than any weakness in the fundamental principles. Nevertheless, when a bonus scheme has gone sour, the easiest answer is to replace it with a different system.

To this end, some companies introduced systems of measured daywork. These took various forms but the basic concept was a time rate of payment linked to an agreed level of performance measured over several weeks or months. However, some firms that abandoned piecework found that they had thrown out the baby with the bath water. The new systems required good management and supervision. Above all, they needed good communications and sound industrial relations.

One of the limitations of many conventional incentive schemes is that they apply only to production workers. Some companies have introduced schemes for indirect workers such as maintenance workers, storekeepers, inspectors, transport workers, etc. A common practice is to base the bonus of the indirect workers on the incentive earnings of the direct workers whom they serve. In recent years some companies have introduced bonus schemes for office workers based on some form of clerical work measurement. These schemes have their merits but the growth of trade unionism in the white collar sector is associated partly with the fear of redundancy and the tighter control arising from bonus schemes.

Another disadvantage of conventional incentive schemes is that they apply to individuals or small groups of employees. Wide differences in earnings between individuals or groups can cause conflict, especially if the high pay is believed not to reflect the effort or contribution of those receiving it. For this reason some companies have introduced company-wide or plant-wide bonus schemes.

Profit sharing is one basis for company-wide bonus schemes. A formula is established for measuring profits on an annual, half-yearly or quarterly basis. An agreed share of the profit is paid to employees in the form of cash or shares. Profit sharing has a history of success in some companies and failure in others. One of the problems is that the idea of enhancing company profits may be anathema or too remote for some employees, especially in a large company. Moreover, inflation has tended to distort profit figures in recent years. This makes for complications in calculating the profit on which the bonus is to be based.

An alternative basis for a company-wide or plant-wide bonus scheme is to link the payroll to sales turnover. But an increase in sales turnover may be associated with an increase in the cost of materials. A better measure of output is added value. Thus a bonus can be devised to link the total payroll to added value.

Such schemes are not panaceas. They have their limitations but they have many advantages over conventional schemes. The principles and practices of such schemes are worthy of further consideration.

7.2 Principle of added value based bonus schemes

The principle behind added value based bonus schemes is really very simple. The past records of the company are studied to ascertain the relationship between added value and employment costs. Using this analysis as a basis for agreement, the company then undertakes to allocate a certain proportion of future added value to a remuneration fund. Thereafter, at regular monthly or perhaps quarterly intervals, the actual remuneration is compared with the remuneration fund. The surplus, if any, then constitutes a bonus fund which is shared out on a pre-agreed basis.

For example, suppose that the analysis of the past records showed that employment costs had represented 60 per cent of the added value. If the company had also been profitable and able to finance its investment programme, it might agree to maintain this ratio of 60 per cent. In future, if the actual payroll is less than 60 per cent of the added value, the company would pay a bonus to make up the difference. The arithmetic is illustrated in Figure 7.1.

Added value is, by definition, the difference between sales income and the cost of materials and purchased services. It follows that in a given situation added value can be increased either by increasing the sales volume or by reducing expenditure on materials and purchased services. If an increase in output can be achieved without an increase in the number of employees, more added value will be generated to enlarge the remuneration fund. If the cost of materials and purchased services can be reduced, more added value will be available to enlarge the remuneration fund. Under the scheme, employees have a vested interest in both increasing output and reducing the costs of materials and purchased services. Moreover, if an employee leaves and the remaining employees can achieve the same output, without replacing the one who left, there will be a bigger bonus fund available to share out.

Such schemes differ completely from conventional individual incentive schemes. The bonus depends not only on the efforts of the individual, but also on achieving cooperation between individuals and groups. Such schemes differ fundamentally from most other group bonus schemes. Bonus is earned not just by increasing the total output, but also by reducing the external costs of materials and purchased services.

The basic principles of all added value based bonus schemes are identical. But there is no universal formula appropriate for all companies. The scheme must be tailored to suit the circumstances. The original Rucker plans were designed to cover shop-floor employees only. Modern added value based bonus schemes tend to cover all the employees in a company or section. There is a whole range of factors to be taken into account in the design of schemes.

7.3 Designing added value based bonus schemes

The success or failure of an added value based bonus scheme does not depend just on the technical design of the scheme. It depends far more on the degree of mutual trust, understanding and co-operation between management and employees at all levels. However, a well designed scheme is much more likely to succeed and stand the test of time than one that is not soundly based.

Ideally, added value based bonus schemes should be introduced

Figure 7.1 Principle of added value based bonus schemes

In the current period		£
Sales turnover	=	1,000,000
Materials and purchased services	=	400,000
ADDED VALUE	=	600,000
Remuneration fund based on		
60% of added value	=	360,000
Actual employment costs	=	330,000
Bonus fund	=	30,000

only where the climate of industrial relations is good. If the climate is poor, it may be possible to improve it by using added value as a communication tool to explain what the business is about and to improve understanding of accounting information.

Assuming that the climate is appropriate, the first step is to study the historical data. This involves a detailed investigation into past records. It is not enough just to examine the annual accounts for the past four or five years, though that may be a good starting point. It is essential to study the pattern of monthly sales, purchases, remuneration and other items.

7.3.1   *Defining added value for bonus scheme purposes*

Although there is broad agreement about most of the constituents of added value, the definition for the purpose of a bonus scheme may vary. The aim of the scheme is to encourage employees to increase output and reduce the costs of materials and purchased services. So it is desirable that employees should feel that they can influence these items.

All the cost items should be carefully scrutinised to ascertain those that can be influenced by employees. These 'controllable' costs should be deducted from sales. But costs which are beyond the control of most employees should be left within the added value. Thus the definition of added value for a bonus scheme may differ from that used in accounting procedures. Instead of using sales minus materials and purchased services it may be better to use sales minus 'controllable' costs. Of course, both sets of figures should take account of stock changes.

Some authorities take the view that costs which vary with volume are 'controllable' whereas costs which do not vary with volume are part of added value. Obviously, in this context, all labour costs are treated as part of added value even though in theory some of them vary with volume. Examples of the kind of expenses that might be treated as part of added value are rent and rates, insurance, advertising and other selling expenses, legal and professional fees, etc. In large organisations the definition of added value at divisional level may include all head office costs, research and development charges, etc., despite the fact that these items include purchased materials and services. For practical purposes, these items are beyond the influence of employees within a division or establishment.

Although there are no hard and fast rules about the definition, there is little doubt that the added value figure should include all remuneration costs, depreciation, interest charges and profit before tax. Some companies have attempted to use a very narrow definition by omitting depreciation and interest charges. Thus their definition of added value amounts to employment costs plus profit. A scheme based on such a definition would be more of a profit-sharing scheme than an added value based bonus scheme. Indeed, in one case, where it was proposed that only half of the bonus fund would be paid to the employees, the formula amounted to a profit-improvement scheme! The definition of added value for bonus scheme purposes should be broadly based rather than narrowed down too much.

7.3.2   Defining employment costs

In the early days of added value based bonus schemes, such as the Rucker plan, the definition of wages was simply the weekly or monthly pay. In those days there was not even the complication of income tax. Nowadays, many employees not only have income tax deducted from their gross pay but also their contributions to national insurance, pension funds, trade unions, welfare schemes, etc. Obviously, the employment costs for bonus purposes must start with the gross pay rather than net pay.

Years ago, the employers' contributions to national insurance were small in relation to the gross pay. But now they are a significant part of the employment costs. They cannot be ignored for bonus purposes. Similarly, the employers' contribution to pension schemes used to affect only a small proportion of the total employees. Nowadays, these costs are significant. They should be included in the employment costs for the purpose of an added value based bonus scheme.

The effect of recent legislation and government policy on wage restraint has been to encourage companies to provide benefits to their employees in kind rather than in cash. The provision of company cars, cheap loans, low-rent housing, etc., has tended to widen the definition of employee benefits. Even the subsidised canteen or company restaurant can make quite a difference to the standard of living of the employee.

Whether these kind of employee benefits, which are also employment costs, should be taken into account in an added value based bonus scheme is open to debate. If the total costs of such benefits are low in relation to gross pay and employers' contributions to national insurance and pensions, they can perhaps be ignored. But if there were a long-term tendency for these kinds of cost to increase, it might be important to take them into account.

The definition of employment costs should be kept as simple as possible, consistent with ensuring that the basis of the bonus scheme is fair and reasonable.

7.3.3 Stock changes in bonus calculations

Calculations of added value should take account of changes in stock levels. In some companies, stock changes are very small in relation to sales and purchases. If so, any minor fluctuations could be ignored. But in other companies there can be very big changes in stock. This is particularly true in industries where the manufacturing cycle time runs into months rather than days or weeks. For example, in shipbuilding, the construction industry, the process plant industry and in heavy engineering, the work-in-progress may change substantially from year to year and certainly from month to month.

If large changes in stocks were ignored, the added value figures and ratios might be distorted to the point of being unreliable. So if the changes are large in relation to the volume of sales and purchases they should be taken into account. This should not be too difficult when dealing with annual figures based on the financial year end. But it may be very difficult, if not impossible, in attempting to derive monthly figures. Yet if the added value calculations were to ignore stock changes, the ratios of added value to sales and of employment costs to added value may vary simply because stock levels fluctuate (see Figure 1.4).

In such cases, there are two possible courses of action. One is to assess stocks in future at the end of each month. This task may not be as formidable as seems at first sight. There is no need to list and value every single item of stock. In many companies, 80 per cent of the total value of the stock is covered by only 20 per cent of the items. If these items are listed and valued, the remainder can be estimated, based on the last reliable stocktaking.

Alternatively, instead of trying to do a monthly check of stock, the changes can be ignored by basing the ratios on moving averages. Instead of comparing one month's employment costs with one month's added value figure, the calculation can be based on three or possibly six months. By this method, the effects of large stock changes on the ratios are reduced. However, the scheme would need to make provisions for adjustment at the end of the financial year or half-year when the actual stock changes are known.

7.3.4 Stability of the ratios

When the definitions of added value and employment costs, and the treatment of stock changes, are agreed, the data for several years can then be examined. An easy way of assimilating the information is to plot the figures and ratios in graphical form. The ratio of employment costs to added value is the key figure. However, it may be desirable also to plot the ratio of added value to gross output or sales and, in addition, to plot stock changes.

Depending upon the size of the company and the complexity of the operations and products, it may also be beneficial to plot some other ratios. For example, instead of treating employment costs as one figure, they can be broken down to show the proportions of wages for shop-floor employees and salaries for other staff. The employment costs can also be split down into gross pay, employers' contributions and other benefits.

Similarly, the added value and employment costs could be analysed in terms of products or departments. This could provide the basis for several different schemes within one large company.

The analysis must be interpreted with care. If the plot of the ratios shows peaks and troughs at certain periods, it may be known that these coincide with booms and slumps in the economy or the industry. The fluctuations may be associated with changes in stock levels. Any noticeable upward or downward trend may be linked with price increases or restrictions or with wage increases or restraints. Usually, there is some rational explanation.

Ideally, an added value based bonus scheme should be founded on stable ratios. But if the analysis shows fluctuations or trends for which the causes cannot be identified, it may still be possible to introduce a scheme. Past history is only a guide to future action. One effect of the scheme is to stabilise ratios which would otherwise fluctuate.

7.3.5 Deciding the target ratios

The analysis of past records provides a basis for setting the target ratios for the future. However, it is not enough simply to draw a line of best fit through a series of points on a graph. It may be necessary to eliminate some of the points as untypical of normal working, e.g. a strike or go-slow would distort the ratios. Similarly, shortages in the supply of materials, causing excessive waiting time, might distort the ratios. Conversely, the effect of very high prices in a boom period might be regarded as abnormal.

However, there are no hard and fast rules about what constitutes an abnormal period. Indeed, it can be argued that the target ratio should be based on the past average, including both booms and slumps. Afer all, the scheme is designed to share out the added value in bad times as well as good. But clearly, it would be wrong to base the future ratio on a past history of unprofitable trading for several years. To take the extreme case, if the recent history showed that the employment costs had exceeded the added value, it would be folly to suggest that the target ratio should be more than 100 per cent!

Even in a company with a good profit record the target ratio is not necessarily the same thing as the historical ratio. The mere fact that in the past the ratio has been, say, 60 per cent, does not mean that it should always be 60 per cent in the future. There may be good reasons for change. The company may be embarking on an investment programme that involves not just replacement of worn-out or obsolete equipment but a change which significantly effects the ratio of capital to manpower. If so, there may be a need to reduce the ratio to, say, 55 per cent or less, in order to pay for the investment.

This does not mean that the employees will be worse off. On the contrary, the effect of the investment should be to raise the added value per head. Instead of getting 60 per cent of, say, £8,000, i.e. £4,800, they may be getting 55 per cent of, say £10,000, i.e. £5,500. However, it may not be easy to explain a reduction in the ratio, or gain acceptance for it. An alternative way of expressing the same thing is to invert the ratio of employment costs to added value. The ratio of added value per £ of employment costs is the same ratio, inverted. Thus, instead of talking of reducing the first ratio from 60 to 55 per cent, we are talking of increasing the ratio of added value per £ of employment costs from £1.67 to £1.82. Psychologically, it is easier to explain the need for an increase in the latter ratio. The extra 15p is to be spent on investment to raise the added value per head and hence the average wage/salary per head.

Of course, the target ratio cannot be fixed for all time. There must be scope within the scheme to review the target ratio from time to time. This could be done, say, once a year when the company is formulating new budgets for the year ahead. The target ratio is not sacrosanct. But it should not be changed without justifiable reason and without prior consultation and agreement.

7.3.6   Calculating the bonus fund

When the target ratio has been decided and agreed, the arithmetic is then straightforward. Each month a calculation is done, as shown in Figure 7.1, to establish the amount of the bonus fund. Depending on circumstances, the calculation can be based on one month's figures or it can be based on a moving average of 3 to 6 months or even longer. However, a moving annual average may tend to stabilise the index too much. It is better to have a measure that reflects the recent performance so that the bonus is linked to recent results.

The calculation can be carried out by the company's own

accounting staff. It involves very little extra work. In a small company with no full-time accounting staff there may be some extra work to arrive at the figure. Even so, the cost of calculation should be more than covered by the benefits of increased productivity. In the early stages of a scheme it may be desirable to have the figures audited by the company's own auditors or some other independent authority. However, experience shows that in successful schemes this independent auditing need not be a permanent feature. One effect of the scheme is to break down barriers of suspicion. The figures on which the calculations are based must be available to representatives of all types of employees. There are no secrets, so no grounds for suspicions.

The presentation of the calculation can vary, depending on whether the target is expressed as the ratio of employment cost to added value or as the index of added value per £ of remuneration cost. The arithmetic is similar in both cases.

7.3.7 Sharing out the bonus

Before the scheme is introduced, there must be an agreed basis for sharing out the bonus. The first question is whether the whole of the bonus fund shall be paid out or only part of it. In some added value based bonus schemes there is no doubt that the employees are entitled to the whole amount of the bonus fund. The argument is that 'the company' also gains because it retains a constant proportion of an increasing amount of added value.

However, in other schemes, the bonus fund is split between the employees and the company. The split may be 50/50 or perhaps 75/25. The argument is that the company needs a bigger share of the added value to pay for additional investment and to finance extra working capital. The effect of such a scheme is that the actual ratio of employment costs to added value will be less than the target ratio. This point is illustrated in Figure 7.2.

This type of bonus scheme seems at first sight to deprive the employees of part of their rightful share of the added value. However, in practice, it has proved acceptable and effective in some companies. The argument that the 'surplus' of added value should be shared between the employees and the company has not been seen as unfair. In any case, what matters is the use to which the company's share is put. If it is spent on extra investment, the main beneficiaries will be the employees who will gain as the added value per head increases.

In effect, this system is one method of reducing the actual ratio of employment costs to added value from an historically high figure

Figure 7.2 Ratio of employment costs to added value in some schemes

In the current period		£
Sales turnover	=	1,000,000
Materials and purchased services	=	400,000
Added value	=	600,000
Remuneration fund based on		
60% of added value	=	360,000
Actual employment costs	=	330,000
Bonus fund	=	30,000
Employees' share, 50%	=	15,000
Company's share, 50%	=	15,000
Actual employment costs including employees' bonus	=	345,000
Actual ratio of employment costs to added value	=	57.5%

to a basis that strengthens the long-term viability of the business.

Irrespective of whether the whole amount of the bonus fund is to go to the employees or only part of it, the second question is whether the agreed amount should be paid out all at once or whether part of it should be put to a reserve, for the benefit of employees, to be paid out in months when the calculation shows no bonus fund. In the example shown in Figure 7.2, it might be agreed that only £10,000 of the £15,000 would be paid out immediately. The remaining £5,000 would be paid into a bank account, controlled by employee representatives, to build up a reserve.

If this system is used, there should be some rules to establish how much should be put to reserve. A simple system is to set a limit to the reserve fund of, say, £100 per employee but with a proviso that not less than half of the bonus should be paid out in any one month. In the early stages of a scheme, it might be better to pay good bonuses to encourage future improvement in performance. Then as the results improve, some of the bonus can be put to reserve for the proverbial rainy day.

Another question that must be settled before the scheme starts is whether certain employees shall not be eligible for bonus. For example, the scheme might stipulate that new employees are not entitled to participate until they have been with the company for three months. Then there is the question of absenteeism. If an employee is absent for, say, two weeks in a month, is he or she entitled to full bonus or only part bonus? The answer might vary with the cause of absence, e.g. sickness, malingering, training, etc.

The basis of sharing out the bonus must also be decided. Is the total amount to be divided by the number of employees so that everybody gets the same share? Or should the total amount be expressed as a percentage of the total remuneration so that the higher-paid people get bigger bonuses than the lower paid? If the latter system is adopted, should overtime premium be included when calculating the percentage? These and similar questions must be settled before the scheme starts. However, they need not be fixed for all time. They can be changed by mutual agreement.

7.4   Installing added value based bonus schemes

The technical design of added value based bonus schemes can vary quite considerably to suit the circumstances, but all the successful schemes have certain features in common. These are more to do with the chemistry of human relations than with the mathematics of financial calculations.

The first essential is the enthusiasm and commitment of top management. The managing director and his senior colleagues must be convinced that an added value based bonus scheme is appropriate and beneficial for the company. Then any problems can and will be overcome. If they are not sure about the virtues of added value, if they are seeking a panacea, if they are simply looking for a new formula for a bonus scheme or a sop to pacify the employees, then attempts to introduce an added value based bonus scheme may simply add to problems.

The whole point about an added value based bonus scheme is that it is not just a system. It is much more a philosophy of management. It is not just a matter of calculating the bonus at the end of the month. It is much more a matter of genuine two-way communication, deliberately seeking the views and ideas of employees to foster their participation and involvement in the running of the business.

Thus, managers who prefer the more autocratic style of management (which may be appropriate for some circumstances) should not rush down the added value road. Once you start down this road it is difficult to go back. Perhaps a better analogy would be to say that it is like plunging into a swimming pool. It can be fatal for the inexperienced swimmer to dive in at the deep end. In added value terms, the deep end means bonus schemes. It is much better to start at the shallow end, learning first how to use added value for measuring output and performance. Once the managers have confidence in this use of added value they can then progress to the middle part, still not out of their depth, by using added value for communication purposes. Experience of this style of management prepares the manager to progress to the deeper water where the benefits can be even greater.

How long should it take to install an added value based bonus scheme? This question is very difficult to answer. In a small business with good industrial relations it could take as little as three or four months. In a larger firm with average industrial relations it could take a year or more. In a giant company with poor industrial relations it could take years. The main reason for the difference is the time needed for explanation, understanding and involvement.

The smaller the number of people within the company or group, the easier it will be to establish and enhance sound communications. In a small business there is less need for formal arrangements, rules and procedure than in the big business. Even so, it is generally advisable to set up some form of committee or group composed of representatives of management and employees. If some kind of joint consultation body already exists it may be possible to use this as the basis of a new body with different functions. Instead of just discussing matters like conditions of employment, working environment, welfare, etc., the new body will discuss wider issues to do with products, processes, materials, customers, suppliers, organisation, costs, prices, profits, investment, etc. However, it should not usurp the functions of any existing body that deals with wage and salary negotiations between the company and the trade union representatives.

The purposes of the committee or council are first to assist in the design and installation of the scheme and then to administer its operation. This does not mean that the committee takes over the function of management. It does mean that management can only manage successfully through gaining the co-operation of the employees. The committee is not there to make day-to-day decisions of management. It is there to discuss and agree issues of importance that affect the performance of the firm. In the final analysis, management must still manage. But it is management by consent rather than management by coercion. It means an open style of management with no secrets.

During the design and installation, the committee provides a platform for exchanging ideas and agreeing action. But the task of communication cannot be left to the committee alone. There must be many other opportunities for explanation and interpretation, for question and answer. This can be done through a series of regular or *ad hoc* meetings in each department or section. This job is much too important to be left to the shop steward or the personnel department. It should involve senior, middle and junior management as well as shop stewards and functional specialists.

'What a waste of time', the autocratic manager might say. Experience shows that it can be time well spent. The time so spent is more than recovered in terms of extra saleable output and lower costs of production. But if the committee proves to be simply a talking shop, or worse still, a nest of vipers, the autocratic manager would be right. There is no guarantee that such bodies work effectively right from the start. But in the modern social climate they are more likely to succeed than the old fashioned methods. There's an old army adage that time spent in reconnaissance is seldom wasted. Properly planned reconnaissance pays off. Similarly, in industry, time spent in communication is seldom wasted. Properly organised, communication pays off.

Of course, there is a vast difference between a committee that simply acts as a platform for talkers and one that deliberately facilitates changes. Meetings can be useful just to provide opportunities for people to let off steam. But all the talking in the world will not of itself increase output or reduce costs. The only way that an added value based bonus scheme can pay off is by people taking deliberate action to change the ratios.

Left to itself, the ratio of payroll to added value will not change much. Indeed, it might well go in the wrong direction so that no bonus is payable. Someone, somewhere in the company must take positive steps to try to change the ratio. This means either increasing the total output, reducing the costs of materials and services, raising the prices, improving the product mix or reducing the payroll. One or more of these must happen if there is to be any bonus.

The initiative for such changes may come from management, especially if it involves new products or capital investment. But the initiative can also come from employees at any level. There may be suggestions for reducing wasted material, increasing output, improving quality, changing designs. Any idea, no matter how small, must be welcomed, evaluated and acted upon. The vital thing is to stimulate a spirit of cooperation and a desire for continual improvement.

Thus the calculation of the monthly figures is not the most important part of a successful added value based bonus scheme. The calculation is important as a record of achievement and as a basis for sharing out the wealth created. But the calculation does not automatically generate a bonus. The bonus arises as a consequence of the action taken to achieve improved results.

Added value based bonus schemes are not panaceas to solve every problem of every business. They tend to be more successful in smaller firms with less than 1000 employees and, better still, less than 500 employees. Some large companies can break down their activities into discrete sections of suitable size for a group scheme. Other large companies find it difficult if not impossible to identify small groups within the organisation. However, this does not mean that they should forget about added value. It may not be appropriate for a bonus scheme but it can be invaluable as a basis for communication and perhaps for wage and salary negotiations.

ADDED VALUE

8

and business policy

8.1   Added value and marketing strategy

It is no exaggeration to say that one of the biggest problems in
Britain, and some other European nations, is that there are too many
people in the wrong business. The industrial revolution started in
Britain. It soon spread to Germany, America and other countries.
But the advantages gained by the pioneers cannot last for ever. The
complex cycle of invention, innovation, production, growth,
development, competition, consolidation, decay and death has
much to do with the added value concept.

Consider, for example, the railway business. Before the invention
of the iron horse most goods and people were moved overland from
place to place either by horsepower or manpower. There was a
limit to the weight and volume of goods that could be moved by a
horse or a man. Although the canals brought a significant improvement in productivity measured in ton/miles per horse, the process
was slow. The stagecoach was another means of improving
productivity beyond the one-man-one-horse index. But again there
was a limit to the number of people that a given number of horses
could convey a given distance.

Suddenly, the locomotive broke through the productivity barrier.
The new railway system could carry more people and goods, faster
and cheaper than the traditional methods. The railway could offer
a service for which people were willing to pay handsomely. The
added value generated per person and per unit of capital employed

in the railways was much higher than in the stagecoach or canal business. As a result, the railways were able to pay good wages to attract good employees. Working for a railway company was regarded as a well paid, secure job. The railways could also pay good dividends to attract plenty of capital. Investors fell over one another to put their money into the new railways.

But competition between the companies forced down the prices that the railways could charge for their services. Moreover, operating costs rose as the railways paid more attention to safety. Eventually the railways generated just enough added value per employee to cover the wages and leave behind a modest return on the investment.

Then along came the motor car and lorry. After several years of development these new vehicles offered significant advantages over the railways. They provided door-to-door service. There was no need to stick to a timetable. Moreover, they could offer lower costs than the railways.

In the face of this competition from a new form of transport, the railways lost traffic and had to cut their costs. The added value they could generate fell. But they had become large organisations, unable to adapt quickly to new circumstances. They were committed with massive fixed assets, unsuitable for other use. They reduced their manpower levels. But they were still unable to generate enough added value to cover the costs of manpower and capital.

Whatever accusations of inefficiency and bureaucracy may be hurled, rightly or wrongly, at British Rail, the fact remains that many railways all over the world have similar problems. They have to be subsidised to bridge the gap between the added value generated and the costs of manpower and capital. In short, they are in the wrong business. No matter how well managed, they cannot raise the added value per employee to a level high enough to cover the costs of both manpower and capital.

A similar story can be told about other industries that, in their heyday, generated high added value per head. Coal mining, ship-building, iron and steel and textiles are just a few examples of industries that have passed their prime. Of course, there are still some businesses in these industries that are doing well, just as there are still some railway lines that pay their way. But many organisations in these industries are not generating enough added value per head to cover all their costs.

However, part of the problem in some of the traditional industries in Britain and other Western nations is that they are over-manned in relation to the capital employed and the level of output. They are facing competition from new plants in the Far East, South America and other places where the manning levels are appropriate to the capital employed. Moreover, some of these countries are paying lower wage rates, e.g. in textiles. But even if their wage rates were lifted to the British level, they could still compete because they are generating more added value per head. In some countries, notably Japan, the wage and salary levels are higher than in Britain. They can afford to pay well because their added value per head is much higher than in Britain.

Thus it behoves every businessman, and for that matter every employee, to ask himself the question: 'Are we in the right business?'. The progressive industries of the last century have gone downhill. The progressive industries of yesterday have one foot in the grave today. The up and coming industries of tomorrow will suffer the same fate eventually.

So which are the industries that should be avoided if possible? And which are the future growth industries? The first question is easier to answer. It is clear that some engineering-based manufacturing industries in Britain and Europe will go the same way as the traditional textile industries. Many companies in mechanical engineering would be well advised to diversify into other sectors. The mass production motor car manufacturers will find it difficult to compete with the low-cost-high-productivity companies in the Far East. The electrical and electronic engineering firms are also facing strong competition from overseas companies which are generating more added value per head.

The industries which have been most successful, so far, in keeping abreast of the competition are chemical- and science-based. By investing part of the added value in research and development they have been able to launch new products at prices that generate higher added value per head.

Of course, this does not mean that every mechanical engineering firm ought to move into chemicals. There will still be scope for ingenuity and skill in engineering to create new products offering high added value per head. And some sectors of the chemical industry may decline and decay. But there is no immutable law that will prevent some engineering firms from going to the wall in the same way that some shipbuilders, textile manufacturers, railways and coalmines have closed down in recent years.

To answer the question about future growth industries is more difficult. Many people have the wisdom of hindsight but few are gifted with foresight. However, a growing number of people are now coming to believe that the future growth lies in services rather than manufacturing. This means not just the well known service industries like transport, communications, tourism, insurance, banking and others. It includes activities like education, health services, and other knowledge-based industries. The fact that many of these latter activities are financed through taxation rather than the market mechanism tends to obscure their potential for generating added value. Britain is no longer the workshop of the world. It could become the service centre of the world. But it can only do so if resources of manpower and capital are diverted from the decaying industries into the sectors with potential to grow and develop high added value per head.

Clearly, these are long-term policy decisions that affect businesses over decades rather than years. So what can a company do in the short term to strengthen its marketing strategy? The answer may lie in an analysis of added value in terms of products, capacities, markets and limiting factors.

8.2 Added value and product analysis

Few firms have only one product. A cement manufacturing business may be the exception. Even so, the single product may be sold in different guises — in bulk or pre-packed. Most firms have several products. Some have hundreds or even thousands of products. Yet many firms do not know how much added value is generated from each different product.

Large firms often have over-complex systems of standard costing. These reveal variances between actual and standard costs but do not identify the added value. The managers are inundated with data about costs but they are starved of the vital information they really need.

Small firms often have quite inadequate costing systems. They use absorption costing principles to cost their products. The overheads are 'absorbed' by being apportioned to each product or job number in proportion to labour costs or machine hours or some other basis.

The result of this misleading system is illustrated in Figure 8.1. At first sight it seems that most of the profit has been made on

Figure 8.1 Product analysis by absorption costing

£'000s	Product			Total
	A	B	C	
Sales revenue	200	200	200	600
Material costs	80	100	120	300
Labour costs	50	50	50	150
Overheads	40	40	40	120
Profit	30	10	-10	30

Product A and that Product C has been sold at a loss. But this is a dangerous way of looking at product costs. If the firm stopped making Product C, the total profit would be lower because the overheads remained fixed.

A better way of looking at product costs is illustrated in Figure 8.2. The material and labour costs have been added together then deducted from the sales revenue to show the 'contribution' to overheads and profit. The biggest contribution comes from Product A but Product C is also generating a contribution. If it were possible

Figure 8.2 Product analysis by marginal costing

£'000s	Product			Total
	A	B	C	
Sales revenue	200	200	200	600
Material costs	80	100	120	300
Labour costs	50	50	50	150
Materials and labour	130	150	170	450
Contribution to overheads and profit	70	50	30	150
Overhead costs	–	–	–	120
Profit	–	–	–	30

to switch the manpower and capital used in making Product C to make more of Product A, there would be a bigger total contribution.

But even this method of analysis may be faulty. It assumes that labour is a variable cost. In theory, this may be valid in the short run. In practice, labour costs are fixed in the short term.

An even better way of looking at product cost information is illustrated in Figure 8.3. This time the material costs, and those parts of the overheads that can be identified with products, are treated as volume-related costs and added together. The sum total is deducted from the sales revenue to show the added value. (The definition of added value used here will be broader than just sales revenue less all materials and purchased services.) It can now be seen that Product A is generating the largest amount of added value in relation to both sales turnover and labour costs.

This example deliberately exaggerates the differences between the products. In practice, the differences may be narrower but still significant. Moreover the example is over-simplified for ease of explanation. In practice, there are more complex factors. For

Figure 8.3 Product analysis by added value

£'000s	Product			Total
	A	B	C	
Sales revenue	200	200	200	600
Material costs	80	100	120	300
Volume-related overheads	20	20	20	60
Total	100	120	140	360
Added value	100	80	60	240
Labour costs	50	50	50	150
Fixed overheads	–	–	–	60
Profit	–	–	–	30

instance, it may not be possible to switch people from producing one product to producing another. There may be highly specialised machinery involved. In some industries, all the output passes through one particular process, e.g. the kiln in a pottery. If so, it is useful to calculate the amount of added value generated per machine hour from the different products passing through the key process.

Another limiting factor may be the market demand. It is no use turning out 50 million pepper pots a week if there is inadequate demand. Thus the answer to product analysis by added value is not necessarily simple. But the exercise is usually well worthwhile.

8.3 Added value and pricing policy

How should product prices be determined? There seem to be at least two schools of thought on this question. Many companies, especially in the engineering industry, price on a cost-plus basis They estimate the cost of materials and labour then add on a percentage for factory overheads. To this manufacturing cost they add on margins to cover administrative and selling expenses then a profit margin. On this basis, the price is quoted to the customer.

In other companies, this procedure would be viewed with horror. They price on what the market will stand. It is argued that prices are determined not by costs but by supply and demand. A new product that has advantages for the customer should be priced higher, even though it may cost less to make.

Which pricing policy is best, pricing on cost-plus or pricing on what the customer is prepared to pay? The added value concept clearly indicates the latter method. Customer satisfaction is measured by added value, not by the costs of manufacturing or supplying a product. Customers pay for the service offered by the manufacturer or retailer in changing the form or location of materials. Added value is not created by incurring costs but by satisfying a demand.

Of course, costs cannot be ignored in pricing policy. If a business is to survive, it must ensure that its income covers all its costs. Pricing at a level insufficient to cover all the costs would be fatal, so managers must know their costs when they are fixing prices. But costs do not determine prices, except in the short term. Rather, it is the other way round. Prices determine the amount of added value left after the business has paid its suppliers. In turn, the added value determines the reward available for manpower and capital.

In theory, in a competitive market, manpower and capital move to the sectors which offer the highest reward. In practice, markets are not truly competitive. There are monopolies and oligopolies of both capital and manpower. The hapless customer is sometimes obliged to pay prices that reflect monopoly pricing policies by capital and/or manpower.

However, over-pricing caused by monopoly should be distinguished from high prices that reflect customer satisfaction. Pricing below cost is folly. Yet pricing above cost but below the market price may also be foolish. Under-pricing may lead to excessive demand. If so, the added value does not fully reflect the satisfaction of the customers.

Some new products are over-priced but too many new products are under-priced. The added value generated covers the costs but it leaves insufficient for further research and development to create new and better products. In the long run, prices based on costs can be the downfall of a business.

8.4 Added value and capital investment policy

8.4.1 Traditional views of capital investment

Over the last 200 years the business world has become more and more capital-intensive. Before the industrial revolution the capital needed to run a business was used mainly to finance stocks and debtors. Some businessmen owned their own land, buildings, ships and other fixed assets. Others paid rent or hire charges to the owners of such assets. But very little capital was needed for machinery because few machines existed.

The businessman, who was more of a merchant than a manufacturer, sought a return on his investment in the form of profit. He hoped to generate sufficient income from trading activities to more than cover the cost of purchases of materials, payment of wages, rent and other costs. The surplus or profit covered the merchant's own living expenses and gave him a return on his investment.

Then along came the inventors with various devices to improve the processes of manufacture. Many of the new machines not only replaced human muscle power with the power of steam, they also produced a better product than could be made by hand. Some machines produced products that could never have been made by hand. But in the main, machines enabled one man to produce what two or more men had produced before. Machinery improved the productivity of labour.

The businessman's view of investment in machinery was quite simple. If the cost of a machine, spread over its working life, was less than the cost of the wages that would otherwise be needed to produce the output, there would be an increase in profit. Capitalists saw machinery in investment terms as a replacement for labour. Profit could be generated either by employing labour or by employing machines. If investment in machinery gave a better profit in relation to the capital employed than using labour for the same purpose, then investment made sense to the businessman.

Thus return on capital was the main criterion in investment policy. Or course, there were the exceptions. Businessmen who were rich enough to afford it could indulge in investment for other reasons such as prestige. And not every investment in equipment paid off. Some of the early Victorians lost fortunes by backing engineers whose optimism exceeded their ability. But, in principle, profit was both the objective and the yardstick of investment.

8.4.2   Recent views of capital investment

The traditional profit-orientated view of capital investment has developed in recent years into more sophisticated forms of analysis The simple concept of capital as a substitute for labour is no longer valid. But present day methods of investment appraisal are not so far removed from the original ideas.

The simplest method of appraising a capital investment project is the 'pay-back period' technique. The expected annual savings in labour and other costs, or the expected additional income, is compared with the proposed capital outlay by dividing the latter into the former. The result is the pay-back period. Thus if a capital investment of £20,000 is expected to result in savings, or extra income, of £4,000 per year, the pay-back period is five years. Given a choice between various projects, the one with the shortest pay-back period is the best investment.

The pay-back method of evaluation is too crude for all but minor projects. A more refined version is the return-on-investment technique. In this method of appraisal, the return expected from the project is compared with a target rate. The project is not accepted if the expected rate of return is less than the target rate. Thus if current returns on safe investment are, say, 10 per cent, the business-man is looking for a higher return to cover the risk involved in buying and installing new plant and equipment.

Even the return-on-investment method is crude. Like the pay-back method it cannot distinguish between projects that pay off early in the life of the asset and those that pay off in later years. When interest rates are high, it is better to go for projects that pay off early than those that pay off later. To overcome this and other short-comings of these methods of appraisal, the experts developed more sophisticated techniques based on discounted cash flows. These methods recognise that £1000 received in five years' time is worth less than £1000 received in two years' time. Inflation and interest rates must be taken into account in capital investment appraisal.

Many articles and several books have been written on capital investment appraisal techniques. The subject has become so complex that large organisations employ full-time experts to evaluate their multi-million pound investment proposals. It is not just that the mathematics are complex. After all, computers can deal with the mathematical problems. There are also many uncertainties such as the effect of extra production on selling prices, material costs, wage rates and other costs. There may be technical problems with new equipment that lead to delay in commissioning or failure to achieve the expected rate of production. There may be changes in corpor-ation tax, capital allowances, investment grants, employment subsidies, etc. All these factors complicate the problems of invest-ment appraisal.

Despite the sophistication of current techniques, all of them suffer from one grave defect. They look at investment purely from a profit point of view. They treat manpower as a factor of production. They ignore the fact that capital investment can and does lead to the opportunity of paying higher wages and salaries than would otherwise be possible.

8.4.3 Future views of capital investment

In the current social climate it is no longer enough to look at capital investment as a means of generating more profit. True, profit

is important both as a reward for risk and as a source of future investment. But what really matters is the amount of added value that
will be generated from the capital investment. Because out of the
added value comes not only profit but wages and salaries. If two
projects offer the same return on investment in conventional terms,
but one generates more added value per employee than the other
project, it should be given preference.

This wider view of capital investment treats manpower not as a
cost to be minimised but as a potential beneficiary from the creation
of wealth. Expenditure on capital equipment can bring not only
bigger profits but also higher wages. Thus employees and their
representatives have a vested interest in the results of capital projects.

Thus the first question in evaluating a capital investment proposal
is to ascertain the amount of added value per employee that the
project is expected to generate. If the proposed figure is higher than
can be obtained from current investments or alternative projects,
then the scheme has passed its first test. The next question is to
decide how much of the extra added value can be afforded to raise
wages and salaries yet still leaving an adequate return on investment
in terms of profit.

Of course, there may be a distinct difference between what can
be afforded and what the company may actually pay. With a militant
group of employees in a situation sensitive to stoppage of work, the
company may be obliged to pay out more than it can really afford.
With a cooperative group of employees willing to look at the facts,
the actual and affordable wage and salary bill can coincide. With an
unscrupulous management, exploiting an unorganised labour force,
the profit may be high because wages and salaries are kept low.
However, there are really two distinct issues. One is the appraisal
of the project in terms of added value per employee. The other is the
question of sharing out the added value. These two issues are confused
if profit is the sole criterion of investment. Each issue should be
looked at separately before the final decision is made.

This means that capital investment is not the sole prerogative of
management as representatives of the legal owners of the business.
Nor is it just the concern of bankers and others who act as inter-
mediaries in putting up the money. It is also the concern of employees
who may be affected by the decision. It is very much their concern
if the investment leads to a reduction in the size of the workforce.
If there is no benefit to the employees from a capital investment
project, it should be no surprise to find that they fail to cooperate.

Thus future investment decisions should take a wider view than
conventional return on capital. They should use added value per
employee as the vital criterion, supplemented with the test of sharing
the added value to give higher wages and salaries and, at the same

time, bigger profits. In the case of very large projects, these may not be the only criteria. The government may take a view that investment in certain industries or certain regions, which may not yield as high a return in added value terms as investment in other industries or regions, may nevertheless be desirable for social reasons. However, there must be a limit to the volume of such investments, otherwise the government may be holding down the standard of living unnecessarily.

The major stumbling block in using added value for investment appraisal is the spectre of unemployment. Employees are interested not just in wages and working conditions but also in job security and satisfaction. But the individual company in a competitive situation cannot afford to ignore investment opportunities that lead to higher added value per head for fewer people. If it holds back, it will become uncompetitive, thus jeopardising all the jobs. Instead, the company should seek to ameliorate the lot of those employees who become redundant through capital investment.

In an expanding business there may be opportunities for employees to transfer from one section that is becoming more capital-intensive to new activities. But not all companies can offer such scope. Inevitably some employees must transfer from one organisation to another. Companies and government agencies can take action to facilitate such transfers. If there is no shortage of well paid jobs in other companies, the process of transfer is automatic. The problems arise in an economy with large numbers of people unemployed and many others under-employed but still on the payroll.

This raises wider issues of the long-term effects of capital investment on employment and living standards. There is little doubt that, over the past 150 years, capital investment has helped to raise living standards of the average British citizen. There has also been a vast change in the distribution of total income, making the poor richer and the rich poorer. But redistribution does not change an average. Only increases in productivity can improve the average standard of living.

In agriculture, Britain has the highest manpower productivity in the world. In 1850, 2 million people were employed in agriculture out of a working population of 9 million. By 1975, the 400,000 people in agriculture produced half the food for a population of 60 million. The increase in manpower productivity can be attributed to capital investment not only in agricultural machinery but also in chemical plant. Of course, investment in equipment is only part of the story. Farmers have also developed better breeds of plants and animals and better methods of utilising equipment and manpower. But the increase in manpower productivity in agriculture has been a major factor in improving the average standard of living.

Similarly, in the basic industries of coalmining, metal manufacture engineering, textiles, etc., there have been substantial increases in manpower productivity over the last 150 years. Again the improvement can be attributed to capital investment, better methods, new products and new materials. In 1850, the mining and manufacturing industries employed 4 million of the 9 million occupied population. By 1975 the mining and manufacturing industries employed 7.8 million of the working population of 26 million. Thus the proportion in mining and industry fell from 45 to 30 per cent. Yet the average standard of living rose thanks to the huge increases in productivity.

If manpower productivity in industry continues to increase it could reach a stage where only 15 or even 5 per cent of the working population is needed to produce most of the manufactured goods we need. Fewer people could produce more industrial added value thus releasing manpower for other purposes. But there would be little point in reducing the numbers employed in industry if those released simply became unemployed. That way, the average standard of living, in a material sense, might rise. But the social problems would be enormous. Man needs the dignity of earning his own living.

The question is whether the people no longer needed in industry can be employed in service activities that generate added value. In particular, there is a need to develop service-based exports to earn the foreign currency to pay for our imports of food. North sea oil will defer the need for such exports but, in the long run, we must either develop more exports or become more self-sufficient.

So major capital investment decisions cannot be viewed purely in terms of profit. They must be viewed rather in terms of their ability to generate high added value per head. And if the added value is exportable or import-saving then the investment can be viewed more favourably than one that reduces exports or increases imports.

But investment for added value alone is not enough. The investor who seeks profit may decide not to invest if wages absorb too high a proportion of the added value. There must be an acceptable balance in the way that the added value is shared out.

8.5   Added value and business ratios

One of the main virtues of added value is its use for business ratios. By monitoring certain key ratios in a business, managers can know more about what is really happening. They are then better equipped to make decisions affecting the future strength and prosperity of the business.

Traditionally, managers have used profit as the yardstick. In particular they have related profit to sales and capital. But profit has become distorted by inflation. What is the use of a yardstick that shows apparent returns of 20 per cent on capital employed which, after adjustment for inflation, are really returns of only 2 per cent? Even without inflation, profit depends on accounting procedures for depreciation, government grants, stock valuation, etc. Profit may be shown before or after interest charges. Above all, high profits may be associated with low wages. Profit alone is not an adequate measure of the performance of a business. So what are the added value ratios that a manager can use to supplement and perhaps replace the traditional measures?

Many different ratios can be developed, literally hundreds of ratios, but only a few are really useful in most circumstances. These can be divided into three categories:

1   Added value related to gross output.
2   Added value related to manpower.
3   Added value related to capital.

Before discussing the use of these ratios, a word of warning is appropriate. The ratios can be used either to compare one business with itself over periods of time or to compare one business with another in the same time period. In both cases, it is vital to compare like with like. The definitions of added value and other terms must be consistent throughout. It is usually easier to ensure consistency within one business over successive periods than between different businesses. Even so, there are traps for the unwary, especially when relating added value to capital.

8.5.1   Added value related to gross output

One of the fundamental factors in any business is the ratio of added value to gross output. In a given type of business, this ratio tends to remain stable despite inflation. Changes in the ratio in a business where the products and markets have not changed substantially indicate either an improvement or decline in the margins available. Other things being equal, an increase in the ratio of added value to gross output means higher rewards for manpower and/or capital.

For simplicity, it may be easier to measure the ratio of added value to sales. However, the ratio of added value to sales can fluctuate simply because of changes in the levels of stocks of finished goods and work-in-progress (see Figure 1.4).

Another alternative is to monitor the ratio of purchases to sales. In effect, added value is the difference between purchases and sales. So a ratio of purchases to sales of, say, 40 per cent would correspond to a ratio of added value to sales of 60 per cent. However, the ratio of purchases to sales can fluctuate simply because the stocks of materials, as well as finished goods and work-in-progress, have changed.

A more reliable alternative is the ratio of added value to purchases adjusted for stock change. This is simply an alternative way of expressing the ratio of added value to gross output. For example, if the latter ratio is 60 per cent, the former ratio would be 1.50 (60 ÷ 40). The advantage of the ratio of added value to gross output is that changes in the ratio can be interpreted more easily. An increase in this ratio from say 60 to 65 per cent is easier to interpret than a change in the ratio of added value to purchases from 1.50 to 1.86.

Comparisons between similar types of business in terms of this ratio can be useful. Differences could be attributed to the prices of the products or the prices paid for materials or to the utilisation of materials and services. The ratio alone will not identify the cause of any difference. But it will point the way to further investigation.

Comparisons between different types of business in terms of this ratio are fraught with danger. For example, two businesses selling a similar product may have very different ratios of added value to gross output. One of them might be simply assembling components made by sub-contractors whereas the other might be making all the components itself from basic raw materials. But if the two businesses have similar operations any difference in the ratio of added value to gross output may be significant.

However, differences in this ratio do have a bearing on cash flow and survival. A business with a low ratio of added value to gross output usually needs more money to finance expansion than a business with a high ratio, assuming similar ratios of stocks, debtors and creditors. It is also more vulnerable to small changes in the cost of materials that bite into the profit portion of the added value. Similarly, it is more sensitive to small changes in selling prices.

Nevertheless, a high ratio of added value to gross output is not necessarily a sign of merit. It may be accompanied by low ratios of added value to manpower and capital. It is more important to seek high ratios of added value to manpower and capital than high ratios of added value to gross output.

8.5.2   Added value related to manpower

A primary ratio of manpower productivity is added value per employee. Strictly speaking, the ratio should be added value per man-hour rather than per man-year. Some businesses have many part-time employees whereas others have only full-time employees. In practice it may be difficult to obtain figures of hours worked, so the simple approach of added value per employee on a man-year or man-month basis is used.

This ratio is fundamental to any business. It limits the amount that can be paid out in wages and salaries per head. No business can afford to pay out more in wages, etc., per employee than it is generating in added value per employee, unless it is subsidised. One of the main objectives of management should be to increase the added value per employee. This will permit the payment of higher wages and other benefits. Indeed, this is the only way that increases in real wages can be sustained.

Comparisons of added value per head within one business over successive periods of time can be complicated by inflation. However, suitable indices can be applied to adjust for inflation. The Retail Price Index is a somewhat crude index for this purpose. It is better to use either the index for Gross Domestic Product at Factor Cost or the appropriate Wholesale Price Indices.

The added value per head ratio can be broken down into categories of manpower. For example, the total added value can be divided by the number of production employees and by the numbers of administrative, clerical and technical employees. Of course, these are not absolute measures of the added value generated by each group of employees. But these indicators are useful for comparing trends within a particular business or for comparing one business with another.

A good example of such ratios is the added value per sales representative. This is a better measure than sales turnover per representative, especially in a business with a wide variety of products or where discounts are offered to customers. All too often, salesmen are paid commission on sales turnover. The salesman is then tempted to sell the products with the lowest margins and the biggest discounts. If the salesman's commission is linked to added value, he thinks twice about offering discounts and selling the low margin products.

Comparison between different firms and industries in terms of added value per head are commonly used for national and international studies of manpower productivity. But such comparisons should not ignore the amount of capital invested per employee. Obviously, a capital-intensive business should produce more added value per head than a labour-intensive business. If it does not, what is the point of the capital investment?

Another useful index is the ratio of added value to wages and salaries. This can be expressed in two forms. One way is to show the wages and salaries and other benefits as a percentage of the added value. This index is useful for wage and salary policy. If this percentage exceeds a certain norm in a given business, there may be insufficient added value left to cover the cost of capital. The norm will vary from a low figure in capital-intensive industries, like chemicals, to a high figure in labour-intensive industries, like clothing. As yet, there is insufficient published data to define the norms but some indication can be derived from *Census of Production* results.

An alternative way of expressing the same index is in terms of added value per £ of wages and salaries. In a business where wages and salaries represent 65 per cent of added value, the added value per £ of wages and salaries would be 1.54. What this means is that for every £1 of wages and salaries paid out, the business has 54p left over to cover depreciation, interest charges, corporation tax, dividends and retained profit. Clearly, if this figure falls below a certain level in a given business, the long-term viability may be affected.

Strictly speaking, it is not correct to talk of the index of added value per £ of wages and salaries as a measure of manpower productivity. This would be true if manpower were regarded purely as a factor of production. From one viewpoint it could be said that what matters is not the number of people employed but the total wage and salary bill. Employing 4000 people at £5000 per year means the same wage and salary bill as employing 5000 people at £4000 a year. But employees are not just hired hands. They are also individuals and consumers. They want a higher standard of living. The best way to achieve this is to increase the added value per head.

Thus is it wrong to speak of an increase in the ratio of added value per £ of wages and salaries as an increase in manpower productivity. An increase in this ratio might be accompanied by an increase in capital intensity which in turn should raise the added value per head. But it could also be associated with a fall in wages and even with a fall in the added value per head.

In practice, the ratio of added value per £ of wages and salaries, and its inverse, the ratio of wages and salaries to added value, tends to vary very little in a given company or industry. Despite inflation, devaluation, the trade cycle, price restrictions, wage controls and other influences, these ratios have stayed remarkably stable.

8.5.3   *Added value related to capital*

Added value can be related to capital in several ways. One way is to express the capital in physical terms such as machine hours or square metres of floorspace. The other methods express the capital in monetary terms.

The physical measures are useful within a given business or for comparisons between similar businesses. The output of a particular plant or piece of equipment can be expressed as added value per machine hour. The higher this figure, the better, other things being equal. These ratios will be affected by inflation, just as added value per head is affected by inflation. Similar indices can be applied to adjust for inflation.

These ratios are particularly useful in a capital-intensive business, especially where the same plant can be used to make different products. For example, in metal manufacture, different grades of material, or different types of plate, sheet, rods, bars, or sections can give very different figures of added value per machine-hour.

The ratio of added value per square metre of floorspace is another useful indicator. It can be used not only in manufacturing business but also in retailing and service trades. Shops and warehouses tend to measure output in terms of sales per square metre of floorspace. But an increase in sales turnover may be accompanied by lower margins. The figure that really matters is the added value per square metre of floorspace.

Relating added value to monetary measures of capital employed is similar in principle to relating profit to capital. Various definitions of capital can be used. One possible measure is total assets, i.e. fixed assets plus net current assets. This is the basis conventionally used for return on capital. But it ignores the other sources of finance such as creditors and short-term loans. A business can show a high return on capital because it is over-borrowing. The ratio of added value to total assets can be high for a similar reason. Whilst this ratio is important for investors, it does not give a true measure of the total assets being used by the business.

A better basis is to relate the added value to the total assets, fixed and current. This ratio can be further split down to relate added value to fixed assets only or to current assets only. Each of these ratios can be further subdivided to look at added value in relation to land and buildings, plant and equipment, stocks, debtors, cash, or other categories of assets.

Care must be taken in comparing these ratios between different businesses. For example, one business might rent its land and buildings and hire its plant and equipment whereas another business owns all its fixed assets. Some small firms undertake subcontract work on a commission basis so they have less money tied up in stocks than a business that finances its own materials. Nevertheless, these ratios can be used to advantage within one business or between business provided that like is compared with like.

Accounting procedures can create problems in relating added value to monetary measures of capital employed. One business might have a high rate of depreciation and very prudent stock valuation policies. Another might have lower rates of depreciation and less conservative views of stock valuation. However, these problems are not peculiar to added value ratios. They also affect ratios of profit to capital.

The main problem with monetary measures of capital is inflation.

If current added value figures are compared with capital figures unadjusted for inflation, the ratios will tend to rise with inflation. Ideally, some form of adjustment for inflation is desirable. It is difficult to prescribe a particular method but any form of adjustment is probably better than none. Again, this problem affects return on capital ratios. One thing is certain. Inflation distorts ratios of profit to capital far more than it distorts ratios of added value to capital.

Finally added value can be related to both manpower and capital. As yet no satisfactory way has been found of adding together units of manpower and capital. This problem, and one answer, is discussed in Chapter 3 and shown in Figure 3.6. Another method is described below.

In theory, a chart can be drawn as in Figure 8.4 relating the capital per employee to the added value per employee. Any business whose ratios lie upon the line AB can be defined as average. But a business like that indicated at point P would be below average because it is showing lower added value per head than other

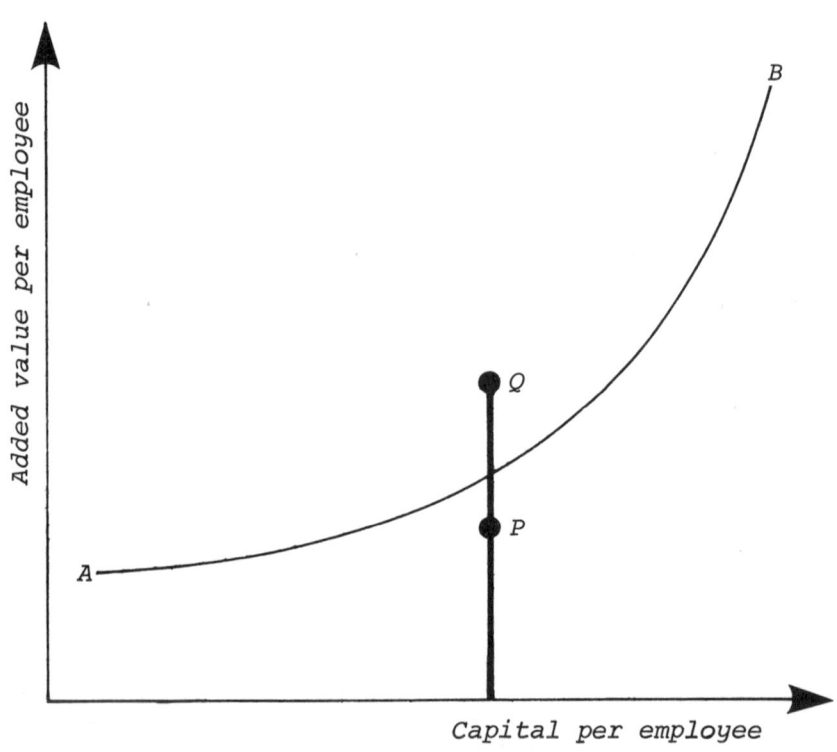

Figure 8.4 Added value related to capital and manpower

Figure 8.5 Net ouput per head and capital intensity.
Source: Census of Production, 1963-75 (All data at 1970 prices)

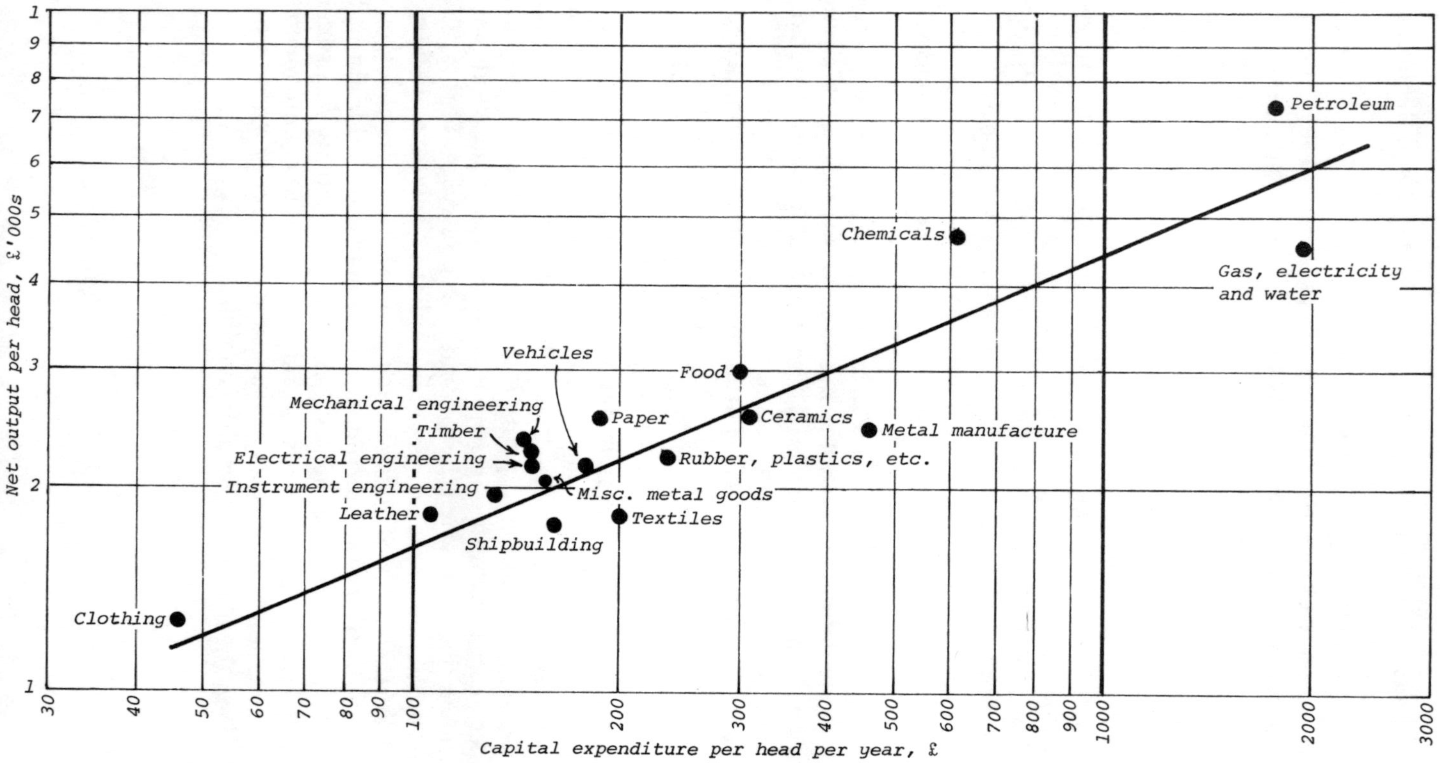

businesses with similar capital per head. Conversely, the business at point Q is better than average.

This simple theory is difficult to translate into practice mainly because of lack of data. Few companies have published sufficient added value figures and there is a lack of consistency in the definition. Moreover, if the capital employed figures were based on published balance sheets, there would be inconsistency in the basis of valuation of assets, etc. However, preliminary study of a number of public companies shows that some firms with high added value per head use less capital than other firms with lower added value per head.

Another source of data is the *Census of Production*. This gives figures of net output per head but not capital employed. However, it does show annual capital expenditure and this can be divided by the number of employees to give an approximate measure of capital intensity. Clearly, the latter figure may fluctuate from year to year but over several years it tends to be higher in capital-intensive industries than in labour-intensive trades.

Figure 8.5 is based on the data for several years. It seems to indicate that the basic theory is valid. But some of the industries are noticeably above or below the average. These figures must be interpreted with caution because net output is not quite the same thing as added value. Nevertheless, there is little doubt that this technique could be refined and developed to provide a basis for comparing the performance of different businesses or different industries.

Comparisons between different countries are even more difficult because of inconsistency in definitions. But most of the studies show that, in certain basic industries, Britain generates less added value per head than other industrial nations, especially the USA, Japan and Germany. Often the disparity is associated with differences in capital per head. But this is not the only explanation. Some studies have shown that even when the amount of capital per head is identical in different countries, Britain produces less added value per head.

Further study of added value ratios, whether between countries and industries or within individual companies, could lead to a better understanding of the factors that affect productivity and standards of living. The main need is to understand and to communicate such information to all employees so that they can play a more positive role in the business of wealth creation.

ADDED VALUE

9

and wealth creation

9.1 Generating added value

Added value is a measure of the wealth created by the application of the knowledge and skills of people to provide various goods and services. These goods and services are generated to satisfy the needs and desires of mankind. Added value measures the satisfaction of customers who buy the goods and services. Added value determines the standard of living. The creation of added value can be accepted as a desirable objective by businesses and other organisations. So what are the measures available for generating and increasing added value?

In terms of a manufacturing business, added value is, by definition, the difference between the sales income, adjusted for stock change, and the cost of materials and purchased services. It follows that added value can be increased either by generating a larger sales income from the existing cost of purchases, or by reducing the cost of purchases for the existing level of sales. Of course, the twin objectives of raising income and reducing costs can be pursued simultaneously.

9.2 Added value and price increases

The easiest way for some companies to increase their sales income is simply to raise their prices. Figure 9.1 shows that a 5 per cent

Figure 9.1 Effect of price increase on added value

£'000s	Present figures	After price increase of	
		5%	10%
Sales (adjusted for stock change)	1000	1050	1100
Materials and services used	500	500	500
Added value	500	550	600
Increase in added value	–	10%	20%

price rise could increase the added value by 10 per cent in a business where added value is 50 per cent of sales. Similarly a 10 per cent price rise could increase the added value by 20 per cent. Of course, if the price increase is required simply to cover the rising cost of materials and purchased services it may not increase the added value. But some firms have used the rising costs of materials as an excuse to raise their own prices by a larger amount.

Even if the higher prices caused some loss of sales volume, the added value might still be higher than before. This effect is illustrated in Figure 9.2. For ease of explanation it is assumed that the costs of materials and purchased services fall in direct proportion to the drop in output. In practice, the fall in costs would not be directly proportional. Nevertheless, there could still be an increase in added value despite the fall in real output.

Obviously, this simple method of increasing the added value of a particular business is a monetary illusion. The employees and shareholders of the business may gain by raising their prices. But their gain is achieved at the expense of the customers of the business, unless they too can pass the higher prices on to their own customers. Somebody in the chain of transactions must stand to lose by virtue of the higher prices. Thus the interests of a particular business may conflict with the interests of other members of the community.

Similar figures could be deduced for a non-manufacturing business. If a hairdresser can get away with a price increase, the customers pay more for the same service. To the extent that the hairdresser gains, the customers lose.

However, the monetary illusion is not as simple as it seems. Some customers may be happier to pay higher prices. Human beings are not always logical in their actions. They often pay higher prices because they think that the product or service must be better than a similar product or service offered at a lower price. As yet, we have no

Figure 9.2 Added value after price increase of 10 per cent and reduction in output

£'000s	Present	Reduction in output of		
		Nil	5%	10%
Sales (adjusted for stock change)	1000	1100	1045	990
Materials and services used	500	500	475	450
Added value	500	600	570	540
Increase in added value	-	20%	14%	8%

better way of measuring the satisfaction of consumers than by the price they are prepared to pay. If customers wish to spend more of their money on certain goods and services and less on other goods and services, the added value of some businesses will increase and the added value of other businesses will decline. Thus the customer determines the added value, not the producer. Nevertheless, some reservation must be expressed about the effects of monopolies, advertising and other means of persuasion.

But there is a way in which higher prices can clearly benefit both the producer and the consumer. That way is through innovation, the introduction of new and better products and services.

9.3 Added value and innovation

When a new product or service comes on to the market to replace an existing product or service, higher prices may be justified if the product or service offers significant advantages to the customer. What matters is not how much the new product or service costs to manufacture or provide but what it is worth to the customer.

A good example can be found in motor car tyres. A new tyre that lasts longer or holds the road better is worth more to the customer. Instead of paying, say, £15 for a tyre that lasts 15,000 miles, the customer benefits by paying £20 for one that lasts 25,000 miles. If, at the same time, road holding is improved, the customer benefits even more.

The new tyre may cost little more than the old tyre in terms of the amount and type of materials used or the time required to manufacture it, but the extra added value generated per tyre is then available to cover the costs of research and development of the new tyre and the costs of any new equipment required. It also offers scope for higher wages and a better reward for the providers of capital.

Of course, if the new tyre lasts longer, customers will need fewer tyres for a given mileage. But this does not mean that the tyre manufacturers sell fewer tyres. The market may be expanding and customers may be encouraged to travel more. And no tyre manufacturer can afford to ignore the competition from other manufacturers. Each must strive to produce a better tyre to attract the customers. Any company that fails to keep up-to-date will go out of business sooner or later.

Thus added value is the spur to innovation. New products offer the producer an opportunity to generate more added value. By satisfying the requirements of the customer, the business generates more added value.

Apart from improving and updating existing products, businesses also introduce completely new products and services that generate added value. Examples include such products as deep freezers, colour televisions, pocket calculators, automatic washing machines and, in the industrial field, computers and xerography. Each of these innovations offers advantages to the customer that help to meet the insatiable wants of mankind. When new products and services first appear on the market, the customer is prepared to pay high prices that yield high added value to the manufacturers and distributors. Then the novelty wears off. Prices fall to the point where the added value is just sufficient to deter manpower and capital from transferring to other activities offering better rewards.

So every organisation, whether an industrial company or a non-profit-seeking service, should seek to develop new products and services that create more added value. The road to higher prices through innovation can benefit both producer and consumer.

9.4   Added value and cost reduction

An alternative method of increasing added value, without changing prices or products, is to reduce the costs of materials and purchased services. The result is illustrated in Figure 9.3. It so happens that the effect of a 10 per cent cost reduction in this business is similar to the effect of a 5 per cent price increase. In another business with material costs running at, say, 80 per cent of the sales turnover, the effect of cost reduction would be more marked. How can a company reduce its cost of materials and purchased services? One way is to negotiate with the suppliers to pay lower prices for materials and services. By this means, a company may benefit its employees and shareholders but only at the expense of the employees and share-

Figure 9.3

£'000s	Present figures	After cost reduction of	
		5%	10%
Sales (adjusted for stock change)	1000	1000	1000
Materials and services used	500	475	450
Added value	500	525	550
Increase in added value	-	5%	10%

holders of its suppliers. In the same way that higher selling prices penalise the customer, lower purchase prices penalise the supplier. Again there is a conflict between the individual company and other people in the community.

In theory, competition should prevent undue reductions in prices because suppliers can refuse orders that do not cover their costs. In practice, large monopolistic customers can squeeze their small-firm suppliers. In doing so, they are acting as monetary bullies. They use their enormous power to gain a higher standard of living at the expense of their weaker brethren. Conversely, small firms purchasing from the large monopolistic suppliers find it difficult or impossible to press for lower prices.

Similarly, the practice of reducing the quality of material in order to reduce costs is a somewhat selfish or short-sighted way of increasing added value. If a motor car tyre wears out more quickly because inferior, cheaper material has been used, the producer may benefit from the higher added value within his company. But his gain is counter-balanced by the customer's loss of satisfaction. In the long run, the manufacturer who skimps on quality will lose customers.

The interplay of such forces is not entirely harmful. The pressure by the customer for lower prices acts as a spur to the supplier to introduce better methods and materials. The right way to increase added value by reducing material costs is again through innovation. Better materials that cost less must be found. For example, plastics may replace metal, particularly where a one-piece moulding can replace a multi-part component.

Another way of achieving genuine and worthwhile reductions in material costs is by minimising scrap and wastage. In many trades, if the customer orders 100 items, the manufacturer buys enough material to make 105 or more to allow for 'inevitable' wastage. The reasons for such losses must be identified, so that action can be taken to reduce avoidable scrap and rejects. Cost reduction then generates more added value.

Sometimes the product or process can be re-designed to reduce material costs. One engineering business found that the weight of a finished item represented only 70 per cent of the weight of the original material from which the product was made. Better methods of manufacture were introduced, giving closer control over the processes to reduce the wastage. The weight of original material was reduced by 15 per cent with no loss of quality or increase in processing cost.

Another example occurred in a quarry where improved methods of crushing reduced the proportion of dust. Thus the costs of material per ton of saleable stone were reduced.

The techniques of value analysis and value engineering are highly appropriate for increasing added value. The organised approach to cost reduction goes beyond conventional methods by questioning the function of the product, component or process. Value analysis may lead to a fundamental re-design of the product. This may be a lengthy process. Nevertheless, it can give spectacular reductions in material costs.

Material-cost reduction need not be confined to raw materials and components. In many businesses, fuel is an important part of material costs. But there may be invisible losses in processes that use steam, electricity or other fuels. Energy saving plays a vital part in increasing added value.

In some organisations the costs of packaging and carriage may offer scope for genuine increases in added value. For example, items sold by the gross (packed 6 X 6 X 4) can be sold in parcels of 150 (packed 5 X 5 X 6) thus saving on the cost of cartons and on carriage if the larger parcel falls within the same price bracket for carriage costs. Methods of packaging can be improved to reduce costs. Shrink wrapping can replace cartons. Light-weight glass bottles can replace heavier containers.

The scope for reducing the costs of materials and purchased services in industry and commerce is enormous. One of the best opportunities lies in the colossal waste of paper in modern offices, especially in the non-marketed sectors of the economy.

A much overlooked method of increasing added value is proper attention to the relative costs of different products. Many companies do not know their product costs accurately. Many companies that think they do know their product costs are often looking at the figures the wrong way. Without reliable information it is all too easy to expand the sales of the products offering the lowest ratio of added value to resources. The right way of evaluating product costs is to relate the added value to the amounts of manpower and capital required to make and sell each product. Then, if the market permits, efforts should be made to expand the sales of the products that generate the largest amount of added value in relation to the inputs of manpower and capital. In turn this policy will enable the company to pay higher wages and salaries and also bigger dividends.

9.5 Added value and productivity

Within one company or organisation, a distinction must be drawn between increasing the total amount of added value and increasing the productivity of resources. The total amount of added value can be increased with no change in productivity. Output can be doubled by using twice as much manpower and capital. The productivity would be unchanged. Indeed, the doubling of output might require more than twice as much manpower and capital. If so, either customers must pay more, to cover the higher costs per unit of output, or else the rewards to manpower and capital must fall.

Thus the mere creation of added value is not enough. Productivity must be increased at the same time. Any increase in added value must be accompanied by a less-than-proportionate increase in one or both of the inputs of manpower and capital. Alternatively, ways must be found of generating the present level of added value by using less manpower and/or capital. The resources no longer required can then be released for use in another company to create more added value.

It is nearly always easier to raise productivity in situations where more output is required from existing resources than in circumstances where fewer resources are needed. The reasons are simple. Human beings resist changes that they feel may be for the worse. No man welcomes the prospect of unemployment or lower pay. Even if the manpower can be reduced, some of the capital inputs may be indivisible. It may be possible to sublet parts of a building that are no longer needed. But idle machinery is a wasted asset.

Thus a high level of demand is a desirable prerequisite for increasing productivity. The overall level of demand in an economy can be affected by government action. But the individual company or organisation is powerless when it comes to converting a slump into a boom. All that the company can do is to try to be in the right market at the right time. No amount of management expertise can create added value from a product or service that people don't want or can't afford to buy.

Assuming that a company is in the right market at the right time, how can its productivity be increased? How can net output be increased without a proportionate increase in the inputs of man-power and capital?

Often the quickest and easiest way to increase productivity is to buy better equipment. Old, obsolete machines can be replaced with modern versions that run faster and produce better quality. A major part of the increase in the standard of living of industrialised nations in the last 150 years can be attributed to more and more mechanisa-tion. The modern loom weaves faster and better quality than its counterpart of 50 years ago. But the loom itself is being superceded by machines for knitting fabrics rather than weaving. In engineering, high-speed tools have replaced slow-cutting equipment. In transport, the modern goods vehicle is much more efficient than the pre-war lorry.

New equipment means capital investment which in turn requires access to adequate funds. Either a company must be generating an adequate flow of funds from its present added value or it must be able to borrow funds against the strength of future ability to pay the interest charges and, if necessary, to repay the capital. But many companies, particularly the smaller firms, lack the financial resources for capital investment.

In any case, the first step should be to ascertain whether existing equipment is being used to the full. Often, plant utilisation can be increased by 20 per cent or more, thus raising productivity. More-over, new equipment should be bought only if its cost advantages are worthwhile, or if it offers a significant change in technology, a process that cannot be carried out by existing machines. Even so, the purchase of new equipment does not of itself increase productivity. Many firms have bought new plant, then failed to make full use of it. Only management action and employee cooperation can ensure that the capital investment pays off.

The alternative to capital expenditure is to find ways of making better use of the present resources of machinery and manpower. This is the purpose of the well established technique of work study. Unfortunately, work study has acquired a mixed reputation. It is sometimes seen as a device for squeezing more work out of the poor hard-worked worker. Some managers view it merely as a means of setting piecework prices. It is true that work study can provide a sound basis for incentive schemes but the improved results should not be ascribed purely to the cash nexus. A bonus scheme is not a substitute for good management. Work study is a far more funda-mental tool than many managers realise. Its aim is to improve the productivity of the resources. It replaces inspired guesswork by systematic analysis of the facts.

Work study, not just on the shop-floor but in the office and even in the boardroom, is a powerful technique for increasing productivity.

It is a foundation stone for other techniques such as production control, costing and budgeting. But there is more to work study than work measurement, popularly associated with stopwatches. Work study incorporates method study with the objective of method improvement. It can lead to substantial increases in productivity, ranging from 10 to 1000 per cent and more. The easy part of method study is devising better ways. The hard part is getting the new ideas willingly accepted and effectively installed. In many organisations, method improvements can be devised 10 times faster than they can be implemented.

Work study, if misapplied, can cost more than it saves. It can upset the relationships between management and employees. But properly applied, it can boost the productivity of both manpower and capital. By generating more added value from existing resources, work study can lead to higher wages and salaries, bigger dividends and better value for the customer.

The creation of added value requires a continual search for new products, new processes and new materials. It means offering the customer more tangible benefits for money. It cannot be done by robbing Peter to pay Paul. Creating added value is a fundamental objective of good management. By creating more added value we can all enjoy a higher standard of living. Thus added value is the key to prosperity.

Reading list

Very few books have been devoted entirely or mainly to added value and its uses. Many books on economics and management mention added value in passing but few devote as much as a chapter to the subject.

Bentley, F.R., *People, Productivity and Progress,* 162 pages, Business Publications (1964).
> Fred Bentley pioneered the 'Share of Production Plan' in Britain. His book is now out of print but can be found in some libraries. Although the examples and figures are somewhat dated, the message is eternally fresh.

Gilchrist, R.R., *Managing for Profit,* 165 pages, George Allen and Unwin (1971).
> Written by another pioneer of the practical application of added value in industry, this book might more appropriately have been titled by its sub-title, *The Added Value Concept.* Well worth reading.

Wood, E.G., *British Industries, A Comparison of Performance,* 462 pages, McGraw-Hill (1976).
> This book summarises the author's research based on the *Census of Production* data. Useful as a work of reference with over 380 pages of computer-printed tables of figures covering 150 British industries from 1963 to 1974.

The following books contain useful sections on added value:

Brech, E.F.L., *The Principles and Practice of Management,* pp66-100, 154-159, Longman (Second edition, 1975).

Gilchrist, R.R., *Works Management in Practice,* pp48-49, Heinemann (1970).

Norman, R.G., and Bahiri, S., *Productivity Measurement and Incentives,* pp18-26, 63-68, Butterworths (1972).

Three organisations have published booklets on added value:

Bentley, D.F., *A Dynamic Pay Policy for Growth,* Bentley Associates Ltd, 55 Dyke Road, Brighton, Sussex.

Frankel, M.R., *Business Performance and Industrial Relations,* Engineering Employers' Federation, Broadway House, Tothill Street, London SW1H 9NW.

Smith, G., *Wealth Creation — The Added Value Concept,* Institute of Practitioners in Work Study, Organisation and Methods, 1 Cecil Court, London Road, Enfield, Middlesex EN2 6DD.

Two organisations have prepared visual aids on added value:

The Company We Keep, a tape-slide programme, S.B. Modules Ltd, 159 Great Portland Street, London W1.

The Path to Prosperity, a film, The Henley Centre for Forecasting, 27 St John's Square, London EC1.

Many articles on added value have been published in various journals and newspapers over the last 30 years. The list below is not intended to be exhaustive. However, it covers most of the more interesting and useful references:

Ball, R.J., 'The use of value added in measuring managerial efficiency', *Business Ratios,* pp5-11 (Summer 1968).

Beattie, D.M., 'Value added and return on capital as measures of managerial efficiency', *Journal of Business Finance,* pp22-28 (Summer 1970).

Broster, E.J., 'Measuring productivity: the delusion of value added', *Certified Accountants Journal,* pp73-80 (February 1971).

Caley, M., 'Plant-wide incentives', *Management Services,* pp16-19 (December 1976).

Cardew, D.M., 'Comparisons of efficiency by the CIR (creativity of income from resources)', *Management Accounting,* pp186-7 (May 1976).

Cox, B., 'Added value and the corporate report', *ibid.,* pp142-146 (April 1976).

Farrell, M.J., 'The measurement of productive efficiency', *Journal of the Royal Statistical Society,* Vol.120, Part III (1957).

Gilchrist, R.R., 'Company appraisal and control by added value', *Certified Accountants Journal,* pp573-580 (October 1970).

Gilchrist, R.R., 'A measure of company efficiency', *European Business,* pp44-50 (April 1970).

Gilchrist, R.R., 'How to measure output', *The Commercial Accountant,* pp139-142 (April 1970).

Gilchrist, R.R., 'Added value as a measure of productivity', *Industry Week,* pp18-19 (9 January 1970).

Gow, E., 'Added value as an incentive', *Factory Management,* pp22-23 (February 1971).

Gow, E., 'Using added value as an incentive to improve productivity', *Accountants Weekly,* pp14-15 (5 March 1976).

Jones, F.E., 'Creation of wealth — comparing attitudes in the UK and Japan', *CBI Review* (September 1975).

Jones, F.E., 'The economic ingredients of industrial success', *Proceedings of the Institution of Mechanical Engineers,* Vol.160, pp115-120 (1976).

Loftus, P.J., 'Labour's share in manufacturing', *Lloyds Bank Review,* pp15-22 (April 1969).

Manklew, J.J., 'An annual wage structure with a high motivation', *Machinery* (14 February 1973).

McCallum, J., 'Productivity and incentives', *The Commercial Accountant,* pp93-102 (April 1970).

Moore, J.G., 'Added value as an index of industrial effectiveness', *Work Study and Management Services,* pp23-46 (January 1973).

Norman, R.G., 'Productivity measurement in manufacturing industries', *Works Management,* pp4-9 (February 1971), pp4-12 (March 1971), pp13-17 (April 1971), pp8-10 (May 1971).

Rothman, J., 'Employees' part of value in car firms', *The Times* (7 October 1975). (Europa VIII-X).

Rucker, A.W., 'Clocks for management control', *Harvard Business Review* (Sept-Oct 1955).

Scott, R.C., 'How to use the value added concept to increase profits', *International Management,* pp75-78 (May 1965).

Sharman, E., 'Added value — a yardstick for business performance', *Business Administration,* pp38-39 (May 1970).

Smallwood, R.E.R., 'Added value per £ of payroll', *Industrial and Commercial Training,* pp366-368 (August 1973).

Smallwood, R.E.R., 'Open the minds with added value', *ibid.,* pp306-309 (August 1976).

Smith, G., 'Adding value for incomes and profit', *ibid.,* pp474-477 (October 1973).

Smith, G., 'Participation through information disclosure', *ibid.,* pp350-354 (August 1974).

Smith, G., 'Effective practical information systems', *ibid.,* pp185-187, 229-232, 331-334, 356-360, 395-400, 447-450 (April-November 1975).

Smith, G., 'Earning value for survival', *Works Management,* pp11-14 (February 1974).

Smith, G., 'Understanding productivity', *Industrial and Commercial Training,* pp301-305 (December 1976).

Smith, G., 'Opening up management style', *ibid.,* pp301-305 (August 1976).

Staples, F.W., 'Added value ratios and productivity deals', *The Accountant,* pp723-5 (7 December 1972).

Swainson, J., 'Learning curve and added value', *The Financial Times,* p15 (30 June 1970).

Swannack, A.R., and Samuel, P.J., 'The added value of men and materials', *Personnel Management,* pp26-29, 41-43 (February 1974).

Thompson, A.G., 'Wealth creation and cooperation — the value of added value', *CBI Review* (Spring 1974).

Thompson, A.G., 'Let the employees see the score', *The Director,* (March 1975).

Wilson, H.A.W., 'Added value in measuring manpower productivity', *Management Accounting,* pp168-170 (June 1971).

Wilson, H.A.V., 'Added value and money averages in company-wide incentive schemes', *ibid.,* pp212-215 (July-August 1972).

Wood, E.G., 'A new light on lame ducks', *Financial Times,* (The Executive's World) (16 May 1972).

Wood, E.G., 'Your productivity slip is showing', *ibid.,* (23 January 1973).

Wood, E.G., 'A better deal for productivity', *ibid.,* (17 October 1973).

Wood, E.G., 'How to add value', *Management Today,* pp73-77 (May 1974).

Wood, E.G., 'The best yardstick for performance', *Financial Times,* (The Executive's World) (4 September 1974).

Wood, E.G., 'Productivity cannot be bought', *ibid.,* (22 October 1974).

Wood, E.G., 'How do you rate in the manpower productivity league?', *Industrial and Commercial Training,* pp369-372 (August 1973).

Wood, E.G., 'Manpower productivity in British industry', *Work Study and Management Services,* pp332-336 (September 1975).

Wood, E.G., 'Stopping the spiral with pay linked to company income', *The Times* (Business News) (10 November 1975).

Wood, E.G., 'A policy for pay and productivity', *ibid.,* (2 and 4 May 1977).

Wren, A., 'Participation through added value', *Industrial and Commercial Training,* pp233-239 (June 1975).

Worth, B., 'Reporting of value added', *The Accountant,* Vol.175, No.5320 (23-30 December 1976).

Index

The index is based on the nine main headings of the table of contents which set out the advantages of Added Value in the principal spheres of running a business: Accounting Information; Bonus Schemes; Business Performance; Communication (at all levels of the business); National Accounting; Policy of the Business; Wages and Salaries; and Wealth Creation. The letter-by-letter system of alphabetisation has been used throughout and the abbreviation 'AV' for Added Value is commonly used.